LETTERS FROM A WARRIOR

P.S. MOM, I LOVE YOU

DALE E. DALLMAN

COPYRIGHT

Letters From A Warrior, P.S. Mom, I Love You
By Dale E. Dallman

Copyright © 2024 by Dale E. Dallman

ISBN: 978-1-958356-40-1

Published by TAWCarlisle Publishing.

All rights reserved. The reproduction, transmission, or utilization of this work in whole in any form by any electronic, mechanical, or other means, now known or hereafter invented, including xerography, photocopying, and recording, or in any information storage or retrieval system, is forbidden without written permission.

For permission, please contact Dale Dallman or TAWCarlisle Publishing at twooden@tawcpublishing.com.

Cover design by Fury Designs

Printed in the United States of America

PREFACE

The faithful heart of every Warrior tells a story. It is a narrative that extends from training to the battlefield. It cries out from the essence of humanity itself. Within these humble pages, you'll find the chronicles of a Marine Corps Warrior who still shows the gentle side of an unbreakable bond with his mother, family, and God.

The faded, ink-stained, original letters might not take you through the horrors of war, but they delve into the depths of the human spirit. Each letter serves as a mirror into the heart and mind of a man grappling with pride, fear, and anger as he longs for home and finds solace in the love of family, country, and self.

As you follow the treacherous path of this Marine Corps Warrior, let us all be reminded of the sacrifices made by our military — our brothers and sisters who stepped forward and willfully volunteered to be among the few, the proud, the UNITED STATES MARINE CORPS!

Remember, only about 200,000 citizens hold that privilege out of 360 million-plus people in this great nation!

It isn't a title you lose; you never "get out" of the Marine Corps. You are a Marine until you draw your last breath.

The Marine brotherhood is not a word; it is a title you earn for life.

Letters From a Warrior, P.S. Mom, I Love You

DEDICATION

This book is dedicated to every mother who received a folded flag in place of her son or daughter. It is dedicated to every Marine who charged up that hill and didn't return. To every service person who returned from Asia only to be spit on at the airport. This book is dedicated to the 56,000-plus warriors who came home in a casket and to the thousands who were MIAs, KIAs, and POWs! The country might have forgotten you, but your brother/sister Marines will still raise a toast in your honor! This is dedicated to a FREE AMERICA!!

CAVEAT EMPTOR

For all formerly active and currently active Marines:

I have posted the letter parts of the book on several Marine Corps sites, for over a decade. I already know the bitches and questions you "crayon eaters" will ask as you read it with a magnifying glass. I know what parts you will question and which parts you will say, "That never happened to me."

"That wasn't the way I remember it."

"We never got a movie."

"It was against the law to hit recruits."

"They used Novocain," and many other "corrections!"

SO—FOR YOU—I am labeling this book "FICTION."

REMEMBER—these letters were written by an eighteen-year-old kid! What he saw through the windshield at that time of his naive life is quite different from what it now looks like from the rearview mirror. Don't make a FUBAR out of it! Read the book; think about the parts that relate to your experiences. If you don't remember Friday Night Smokers, it's ok; some will. If you didn't do squat jumps (they are different than squat thrusts), it's ok; some did. You all know that any of the "physical adjustments" were meant to create a Marine. Some drill instructors abused the system, and some delighted in harassment.

BUT—I wouldn't change any of it! It broke me down to a manageable zero and re-built me into a UNITED STATES MARINE! It saved my life, and in later years, it made me "IMPROVISE, ADAPT, and OVERCOME.

RELAX, enjoy the book, or—if it just gets in your craw to the point of creating ulcers—throw it in the "burn barrels!" Or—better yet, donate it to your local library! SEMPER FI!!!!

The majority of the people mentioned in this book are now deceased. Many failed Bootcamp, and many were out of the country. Quite a few are on the WALL, and quite a few are solid businessmen. I sat with two of them and reminisced as their lives ended. We recounted experiences, told stories, and I watched as their lights went out. I have used some of those conversations to put together parts of the stories. I'm sure some of this is embellished, but I'm sure some is also dead on. You be the judge.

BUT—if the book makes you think, if it makes you remember, if it makes a laugh or tear come to you, I have accomplished what I set out to do!

Letters From a Warrior, P.S. Mom, I Love You

TABLE OF CONTENTS

LETTERS FROM A WARRIOR	
COPYRIGHT	
PREFACE	
DEDICATION	
CAVEAT EMPTOR	
MILITARY LETTERS	1
CHAPTER ONE	4
CHAPTER TWO	23
LETTER #1	35
CHAPTER THREE	37
LETTER #2	53
CHAPTER FOUR	55
LETTER #3	68
CHAPTER FIVE	70
LETTER #4	78
LETTER #5	87
CHAPTER SIX	89
LETTER #6	89
LETTER #7	92
LETTER # 8	94
LETTER # 9	97
Dreaming	98
LETTER # 10	102
CHAPTER SEVEN	104
LETTER # 11	105
LETTER #12	106

LETTER # 13	108
CHAPTER EIGHT	110
LETTER #17	110
LETTER #18	111
LETTER #19	116
LETTER #20	120
CHAPTER NINE	122
LETTER # 21	124
CHAPTER TEN	126
LETTER # 22	128
CHAPTER ELEVEN	130
LETTER #23	132
LETTER #24	135
CHAPTER TWELVE	137
LETTER #25	140
LETTER # 26	141
CHAPTER THIRTEEN	144
LETTER # 27	146
LETTER #28	148
CHAPTER FOURTEEN	150
LETTER #29	152
LETTER #30	153
LETTER # 31	157
EPILOGUE	163
U.S. MARINE CORPS	166
ABOUT THE AUTHOR	167

MILITARY LETTERS

I'm sharing these letters that were sent home to family and friends during my days at Boot Camp and early station placement with you to give you insight into the daily life of the United States servicemen and women. My time of action, which not only placed me in combat, but in a recovery room after receiving wounds from enemy fire, was during our nation's time as "OBSERVERS."

The official US policy at that time was no combat and no armed Americans in Vietnam!

DATE LINE: Minot Daily News
Friday, September 23, 1960

```
Twenty-eight   area   volunteers   for   the   United
States  Marine  Corps  have  signed  up  at  the  Minot
```

recruiting office located in the post office with Sergeant Jamie D Weets as part of the North Dakota Platoon. They will leave Fargo, North Dakota on October 28 for training together at San Diego, California.

Sergeant Weets said others interested in joining will have until October 26th to volunteer with the North Dakota Platoon. Those already enlisting are Allen Harstad, Vernon Reiter, Dale Dallman, Vacky Hambeck, Richard Shervold, Phil Aus and Virgil Hanson, all current residents of Minot.

The group enlisting at Minot will leave the Magic City on September 26 by train for Fargo, North Dakota and will be processed the following day. The entire North Dakota group will be guests at a luncheon in Fargo, which will be attended by Miss North Dakota, Carol Olson, as well as other dignitaries. On October 28th, the group will go by bus to Minneapolis, and from there, by plane to San Diego.

Minot Savings & Loan has donated trophies, which will go to the outstanding Marine in the North Dakota Platoon and to the top marksman during the recruit training time.

Organization of the North Dakota Platoon was under the direction of Major D.A. Chapatti of Minneapolis. Captain J.A. Van Der Elzen of the project office expressed surprise at the progress that already has been made on the North Dakota organization.

This was the beginning of a trip that would prove to be funny, frightening, deadly, and always interesting. The group of farm boys that set out that day had no way of knowing how green and gullible they were.

The upcoming letters show just how our unsophisticated and naïve perspectives were changed as we entered into our new surroundings and how we reacted to the Marine Corps.

CHAPTER ONE

Minot, North Dakota was a small town, feebly sustained by the railroad, farming, and an Air Force Base. It was a prairie town that was built by stout German, Swedish, and Norwegian immigrants, with a focus on family, ethnic, and occupational unity.

Many generations of offspring lived only miles apart, with the majority of the population never traveling more than a few hundred miles from the original farmstead. Outside influence was limited to movies on Saturday nights and the few books available at the drugstore counter.

The flat, cold prairie stretched endlessly in front and behind as far as the eye can see, dotted only by a few trees, an afterthought of Mother Nature amidst the bleak landscape she created.

Winter was a case of survival. The temperature could drop to 60° below zero and stay there for weeks. Snowfall could pile up in 6-foot drifts, and the ground blizzards would seem to go on endlessly. Due to the zero visibility from the blinding snow, the farmers often ran ropes from the house to the barn, so they could get back and forth to feed and water the livestock.

Many a person lost his or her way in the harsh conditions, lying in the frozen coffin of ice and snow until June, when finally, the spring thaw revealed the gruesome evidence of demise. They all had wondered aimlessly in circles, seeking shelter until the final moments before death. It was then that they started feeling a certain warmth that caused them to strip off their clothes to lie down, basking in the deceptive heat of the snowstorm.

Summer was just as harsh, and although short-lived, it wore on the souls of the young and old alike. The spring rains made planting an agony that a person could barely endure. The summer heat scorched the crops that mankind slaved to cultivate. The days brought the temperature up to 100 degrees, and the nights were invaded by swarms of mosquitoes that seemed capable of draining your blood in minutes. Farm labor was only one step above what Lincoln fought a war to stop. The only salvation was to go broke and be forced to move to town to work for wages.

This is where I grew up. My father was an engineer on the Great Northern Railroad. Don and Inga Dallman, myself, two brothers, and Grandma Dallman occupied a one-bedroom house with an outside toilet on the north side of Roosevelt Park. The area was unique because the actual pens for the animals surrounded the two-story house.

The elk pen was right outside my window. The night air was constantly pierced by the bugle of elk, the roar of lions, and the screams of a black leopard. Ducks, geese, and peacocks wandered in and through our yard as if *we* were the intruders. We were constantly digging out badgers, prairie dogs, and rabbits that moved into our yard. The dog always smelled like a skunk, and the black bears escaped so many times, that we named our front yard tree, "Bear Hollow." A wild beehive in the tree always attracted the bears, and we were constantly being awakened by the park rangers as they captured the old sow and her cubs, returning them to the confines of an escapable cell.

Minot's Main Street runs north and south, with the Great Northern Roundhouse and workshops at one end and the largest hospital in the area at the other. The town was the star of the state at one time. During Prohibition, it was known as "Little Chicago" because of the whiskey smuggled from Canada passing through Minot! Fine buildings were constructed to survive. Brick and concrete were hauled many miles to provide permanence and protection from nature.

Ellison's department store was the cornerstone, and the rest evolved in straight patterns, block after block. Piggly Wiggly was the dominant grocery store, with Minot Drug, Barness Bakery, and JCPenney all claiming their share of the traffic.

Minot was proud of its schools. Nobody had to walk more than one mile to reach his or her school. The only high school was in the center of town, a sprawling red, brick building that covered an entire block. It featured a swimming pool and an enclosed basketball court. The grade schools were also solid brick and intimidating. Sunnyside Grade School left a lasting impression on my life.

I still vividly remember the most embarrassing moment in my young life. The entire class was choosing a name for the basketball team, and we all were asked to give suggestions to the teacher. I raised my hand and told her I would like to suggest "Soo." She wrote "Sioux" on the board, and I quickly raised my hand again to tell her I had said "Soo." She and the whole class laughed as they realized the only way I had ever seen the spelling of "Sioux", was on the boxcars rolling through town on the Sioux "Soo Line" Railroad!

The town is snuggled in the twisted basin of the Mouse River, as it is known locally, or the Souris River, as the maps indicate. The only trees are found right in the basin, and the few there are nurtured in the unfenced yards. Sidewalks are a luxury, and many of the streets are gravel.

Bars dominated the existing neighborhood businesses, even though North Dakota was a "Blue Law" state. The entire town was closed on Sundays, no business life at all. You were expected to spend Sunday as a family day.

My buddies and I knew somewhere out in the big world, people were having fun and leading exciting lives. Our days consisted of cruising in front of the YMCA and parking in front of the Dutch Mill Drive-In. We knew this couldn't be the high point of one's life.

Those who had escaped Minot would sometimes return, bringing news of life outside the confines of North Dakota. The descriptions of Minneapolis, Denver, Omaha, and even the ocean—from one person who had actually gone to California—were fantasies we couldn't grasp. The conversation involving a 30-story building was muddled by our inability to fathom building heights greater than that of a grain elevator.

We all read Life Magazine and Playboy if the soda jerk didn't catch us, and we had almost seen bare breasts. We passed around the torn-out pages of brassiere ads from the "Monkey Ward," "Sears & Sawbuck," and JCPenney catalogs. We would hold group fantasy sessions behind the school as we looked at the wrinkled pages. Life would be wild if we could just escape the flatlands of North Dakota.

We all knew that the real people of the United States didn't live like us. We knew that California was a land of plenty and that if we could escape, we would find gold at the end of that rainbow. The girls pictured in Playboy beckoned to us, "Come to California. We are waiting to grant your most secret desire."

One of our members, Todd, claimed he had sex with Mindy Chambers behind the school. We all laughed and claimed it as BS, but each of us asked him privately to describe what happened.

Todd told me that it happened really quickly. Mindy teased him and called him a cherry. She dared him to open his pants and show her his privates. Todd told her he would if she let him see her breasts. She laughed and opened her shirt. Mindy had extremely large breasts and was the star of all the boys' dreams in school. (I remember a few private fantasy sessions with her breasts as the highlight.) Todd said he reached over and touched one through her bra. Mindy just smiled. He grew braver and used both hands in a massaging action. She told him, "Slow and easy." Todd said the nipples began to poke his fingers, and he worked the bra up over one. The sight of the bare breast made him gasp, and he gingerly placed the nipple between his fingertips. Mindy sighed and moved closer. She ran her hand down the front of his pants. His reaction made her laugh.

She slowly unbuttoned his Levi's, and they dropped to the ground. The shorts were much more complicated, with her trying to disengage him from the otherwise usually convenient little hole in front. Todd said she grew quiet as the engorged member sprang free and began jumping in the breeze!

It being her first time up close and personal with that part of the male anatomy, Mandy murmured, "My God," when she saw the size!

She touched it and nervously laughed as it jerked in her hand. She slowly massaged it and drew closer to examine it more thoroughly in the sunlight. She dropped to her knees to get a closer look as she continued massaging, causing moisture to appear.

She wiped it off and said in a husky voice, "Does it feel good?" She continued her ministrations, watching intently as he jerked in her hand. The groan Todd let out made her laugh. She continued to stroke the shaft and watched intently. The excitement was too much for a young virgin boy. The sudden ejaculation shot all over her face and hands. She screamed and fell over backward! This brought other students around the corner! Mindy staggered to her feet, closed her shirt and glared at Todd and the other students. Todd just stood there, his shorts at his knees, and the evidence of what had been happening on the ground in front of him.

He just kept saying, "I'm sorry," as she grabbed her books and ran away.

He told me that counted, that he could say he had busted his cherry on a girl.

He never could get Mindy to talk to him again. She must have shared something though because the girls paid more attention to Todd after that. The nervous giggles and the brazen behavior they displayed in his presence went without saying. They were curious about the adventure that Mindy had experienced! Todd became very popular that summer.

The Dutch Mill Drive-In was our headquarters. We ran the town, or so we liked to think. Our gang consisted of ten or twelve hard-core members and another 10 or so tag-along, sometime members. We grew up in the Elvis Presley afterglow era. The uniform of the day was t-shirts, cigarettes (Camels), rolled up in the sleeves, black leather jackets, Levis, and black engineer boots. The hair was either a crew cut or a long Vitalis greasy look in a Ducktail.

The cars were chopped, lowered, skirted, with all chrome removed, and leaded with a sun visor. The door handles were buttons hidden in unlikely places like the grille. A lot of us cut the bottom half of the steering wheel off. The implication was to make it easier for sex.

Uncle Sam made the mistake (in our opinion) of planting an Air Force Base twelve or fifteen miles north of town. It was a huge base that was home to the B-52 Bomber and the Minute Man Missile Wing. We hated the Air Force flyboys. We delighted in causing them as much discomfort as possible. We could not tolerate them dating any of the local girls. We had our honor at stake. It was our duty to protect all the sisters and daughters from the evil hoards. Secretly, we figured they had to be doing what we wanted to do, and the jealousy was boiling in all of us.

This all led to an incident that caused one of the biggest fights in the history of Minot. When Todd found out that Mindy was going out with a fly boy, we searched the town from one end to the other. Carloads of us, one car behind the other, were running up and down the streets looking to serve justice on the out-of-state ass wipes that dared to violate our women.

We found them parked up on North Hill. The convoy of cars surrounded the old beat up vehicle he was driving, and tempers boiled as we saw both rise up from God knows what they were doing and peak out the window.

Todd was the leader of the pack that night. He had to protect his "first." About ten of us walked up to the driver's door. Todd jerked the door open, and all of us stood in awe as the light hit Mindy's monster bare breasts. Still to this day, I can close my eyes and see those perfect glands, the nipple standing out against the sudden cold air and the look of defiance in her eyes. She did nothing to hide herself from our view. She slowly turned and asked what the hell we thought we were doing. Nobody said a word. The whole group was caught in a time warp by those amazon breasts. Todd seemed to come to his senses first. He shouted for her to get dressed and grabbed the fly *face* by the shirt and yanked him out of the car.

The fly *face* tried to stand up, but his pants were at his knees with the belt caught on the seat. It tripped him up and when he did stand up, the black dirt that clung to his genital area was glistening with moisture. We all knew instantly what had been going on when we arrived!

Todd became enraged and kicked the fly *face*. He hit the ground, and all of us joined in. Each one of us struck blows. We all were punishing the Flyboy out of jealousy. He was receiving from Mindy what we had only fantasized about, and we were angry. The beating continued until all of us were exhausted.

The poor Flyboy was rolling on the ground and moaning as we all paid attention to Mindy, who had exited the car. The fact that she was still bare-breasted stunned us into complete silence. She walked over to the Flyboy and spoke to him as if he was just sitting there enjoying an after-the-fact cigarette.

She told him he wasn't her type and not to call anymore. She turned to us and gave the perfect eunuch look as she told all of us, we didn't even have the right to look at her. She reached inside the flyboy's car, pulled out her bra and shirt, slowly dressed and walked down the hill. We stood brain-dead until the Flyboy finally raised to his knees. Squealing tires and flying butts were the order of the day as we ran from our shame and fright.

A week later, The YMCA was the scene where we took our recompense, as 200 Air Force surrounded us! The beating we took was long and mean. The last thing I remember before I went unconscious was the Flyboy, who had been with Mindy, standing above Todd and urinating on his face.

Life slowly came back to normal. The wounds we suffered were only superficial. The biggest wound was dealt to our pride. The Flyboys had come into our town and belittled us in front of the locals. We were the ones the cops warned to cool it. We were the ones that were blamed for causing the trouble. Nobody understood that we were only protecting the virgins in Minot. We should've been awarded a plaque or something!

It was Adam that came up with the solution to our plight. "We should all join the Marine Corps." He had read in life magazine that they were the elite of all fighting forces in the world. The Marines were feared by everyone. The magazine went on to list all the battles where the Marines fought hand-to-hand and won against overwhelming odds. The book named the heroes, Chesty Puller and others, who had the admiration of the world leader. The book told how even the Green Berets and the Navy Seals were frightened to death of the Marine Force Recon units. The book went on to explain that the Marines were looking for "A FEW GOOD MEN." Adam said he had called the Marine recruiter and had an appointment to talk to him at the post office the next day. He wanted all of us to go together. We all agreed to accompany him.

The meeting went well. The Sergeant told us how wonderful boot camp was. He told us that we would have a ball. In fact, he told us everything we wanted to hear.

He had been to California and met the women in the playboy magazine. He even had sex with Ms. June. The Sergeant told us that the Marine Corps uniform was an instant ticket to any woman we wanted and said we didn't even have to ask. The woman would pull us off the street.

He also told us the Marine Corps would furnish us with free beer and all the food we could eat. We only had to answer to our sergeant. We could go to exotic countries, and the Marine Corps would pay for it all. We could get away from Minot and be taught to kill at the same time.

Looking back, I've noticed that juvenile minds seem to be very short-sighted. They only see what is immediately bothering them. Being that the Air Force was after us, the thought of a four-year commitment didn't even register.

I just couldn't understand why my father didn't think it was a good idea. He preached about college, the future, hazards, ideals, and family. My family wasn't rich, but my father was dead set against any Dallman being in the military. This went back to World War II and the fact that my uncle, Ivan Dallmann, was a Nazi SS officer that went into the bunker with Adolf Hitler and never came out.

It seems my family had immigrated to get away from the

shame of my ancestors and to have their children born in a free country. I thought he was dumb as a post and was just trying to control me.

I met with the same resistance that the rest of my friends met with. We spent many a night drinking Grain belt beer and comparing how dumb our parents were. We somehow had to convince them that destiny was calling.

The world was passing us by, and it was also more than obvious that the United States of America was not safe unless we were the protectors. We owed it to ourselves to find a way to show our parents how wrong they were and how committed we were to serving our country (by getting out of Minot).

Vacky came up with the answer. We all knew that the judge in Ward County was a hard old cuss. He sometimes gave offenders the option of reform school or military. We just had to commit a crime that would land us in front of the judge. Though we contemplated everything from robbery to murder, we settled on stealing a car and driving to Montana. We had been told that the State Line Club had topless dancers and didn't check ID.

Gronvold Blessener Motors was to be our target!

Todd, Adam, Thomas, and I were going to commit the actual theft. The plan was for us to look at the cars while one of us stole a set of keys. It went perfectly. I pocketed a set of keys for a big white Oldsmobile, and we wandered around all night drinking Pabst and working up our courage.

When we were finally ready, we pulled up behind the

dealership, and I crawled to the car I had the matching keys to. I reached up and unlocked the door. I started to open it, but suddenly, the sheriff was standing on my other hand. He asked me what I was doing. I wet my pants! It seems we didn't fool anyone. They knew we had the keys and were waiting for us.

My father was sitting at the jail as I was brought in. I spilled my guts, blabbering that we thought this was the best way to convince everyone that the only choice was the Marine Corps.

A meeting with all our fathers was held. The topic was the stupidity of their sons, and the outcome was a unanimous vote to allow us to have our wish. We could have our medicine since we were bent on destroying our futures. My father gave me his consent but told me I wasn't his son anymore!

We couldn't believe our ears. We could go. The world was our oyster. We called the recruiter, and eight of us signed up. The rest of our gang that wasn't allowed by their parents to join were chastised and kicked out of the Dutch Mill. In our eyes, they were cowards. They could stay in Minot and rot for all we cared.

We had so much planning to do, and the time was so short. We only had a little over a month before we left. We had to do everything we had been wanting to do. The problem was we didn't know what we wanted to do.

We drove all over trying to impress people with our decision. The young girls would giggle, and they actually paid more attention to us. Some who only laughed at us before, were now willing to party. We were riding high. We planned parties at an old, abandoned farmhouse and the eight of us were surrounded by girls. The beer flowed, and the cherries popped. Life was good.

We traveled all over the state. We invaded small towns and battled for the local girls. The news that we had joined the Marines traveled ahead of us. We were a force to be reckoned with. The local boys would stand back as we swaggered and teased our way into the heart, and sometimes the pants, of the local damsels. We thought we were invincible. No one could tell us we didn't own the world.

We had a flat tire in one little town in southeast North Dakota. I pulled right up to a car like mine on the street and started to remove the tire. The local Marshall came roaring up to arrest us for stealing. I not only talked my way out of the arrest, but I convinced him to help hold the jack while I proceeded to steal the tire.

I told him the story of our destiny. The Marines wanted us, and we would be responsible for the lives of him and his children. We would not be held responsible for the petty crimes of other mere mortals. We bid him farewell, slipped him twenty dollars, and roared down the road. We did not see the smile and the look of knowing something we didn't on his face.

The one place we walked the narrow line was in Minot. The Air Force was still on the prod. We had a few run-ins with them since the big fight, and we hurled insults and threats but avoided contact. The memory of our beating was still fresh in our minds.

Todd and I decided to get out of town and go hunting. The

only season open was bowhunting. We didn't know anything about bows and arrows but figured it couldn't be too hard.

We went down to a sporting goods store named Trader Dans. It was located right next to Dakota Drug. Dan met us at the door and proceeded to equip us with the best in archery equipment. When it came time for the bill, he asked how we planned to pay. Todd said, "Charge it."

Dan asked Todd what he did for a job, and Todd told him he had just joined the Marine Corps. Dan told him sorry, but he would need a cosigner. I told Dan I would cosign for Todd. Dan said, "Okay." The same scenario took place when I needed to pay.

Todd cosigned for me, and we left with our new equipment in hand and Trader Dan scratching his head. We immediately went to my house to practice. Living in the Zoo area gave us ample targets! Todd quickly killed the neighbor's pet rabbit, and I think we might've gotten a cat or two.

The two of us drove to Garrison Dam and walked into the trees below the dam. We spent some time looking for deer but boredom quickly took over. We drove to Garrison and traded our new gear for beer and whiskey.

We made one quick call to Minot, and the next three days were spent at the farmhouse partying. I was arrested for running down the road buck naked, chasing some sweet young thing that had my pants.

Todd and I made a vow to my mother that we would stay home and be good if she would put up the bail money and get us out of jail. Mom and my older brother drove to Garrison and brought us home. My father had fully disowned me by this time.

We were perfect house guests for a week. We did chores, ran errands, and proved that we could be trusted. Mom was so happy that she decided to give herself a birthday party. She invited all her friends and a few of ours. The day dawned a perfect fall day in North Dakota. The air had a crisp, fresh scent. The sun was full and promising warmth later for the joyful event. Mom was in a festive mood. She baked a cake, prepared the meal, and was all ready until she realized she didn't have any soft drinks. She asked Todd and me to go to the Piggly Wiggly and get a couple of six-packs of 7-Up. Dad had just been paid, and the only money she had was a $100 bill he gave her for groceries. She told us to hurry and gave us the hundred.

We called her collect from Forman, North Dakota three days later. The sheriff told her we were being held for influencing a minor. It seemed we had picked on the mayor's daughter who was only sixteen. The mayor wanted us tried and hanged! Mom and my brother again drove to pick us up. I couldn't look her in the eye, but Todd told her we had the soda pop in the car! All she had to do was pay the impound fee. My old car is probably still sitting in the impound lot in Foreman.

The time to go was finally upon us. We all boarded a bus for Fargo and were taken to a swearing-in ceremony.

Miss North Dakota was there, and a host of local politicians wanted their pictures taken with us. We were now known as the

"North Dakota Platoon."

The new platoon's next stop was Minneapolis, Minnesota. Here we had to take physicals and be processed. They put us up in a hotel in the bad part of town, and we were allowed to go out each night after processing. We were totally in awe! The size of Minneapolis was mind-boggling. The buildings were beyond our imagination, and the sheer volume and diversity of people on the streets was frightening but exciting!

We couldn't get in any of the bars, so we wandered the streets. We saw police cars with their sirens blaring everywhere. The bums in the street kept hitting us up for money. We stopped by a group of ladies on a street corner, and they asked us if we wanted to party. We said, "Sure."

The first one approached me and asked if we had rooms. I told her sure we did, but my mind was racing. She was the first black person I had ever seen, and as my eyes adjusted, I saw she was wearing a short short skirt. The blouse she had on was open, exposing more of her breasts to me with every movement.

She looked at my face and laughed and stepped closer saying, "Here honey, let me show you something." She took my hand and placed it between her legs. She had no panties on and felt the course hair as my hand disappeared. I jerked back like I had been burned. She laughed and told me to call her when I grew up.

The women all moved down the street laughing and making fun of us "honky" boys. We all just stood there on the street with the vision of the black girls rolling around our minds like an omen of things to come. The world was here, but all of a sudden, we weren't as ready as we'd once thought. We found our way back to our rooms and spent the rest of the night drinking soda and lighting farts.

The fact that we belonged to the government still hadn't set in. We were loaded onto a plane, and our trip to San Diego began.

None of us had ever been near an airplane, let alone on one, in our whole lives. The fact that we were about to take off into the air did not compute. We all found a seat and kept laughing and joking. Other people on the plane looked at us with distaste and contempt as we continued to try to prove how macho and superior we were to everyone else.

The start of the engines was the first indication that we might not know as much as we thought we did. The plane taxied from where we boarded, and the pilot started the engines. I suddenly became cautious and started to pay attention. The sides of the plane vibrated, and the power of the engines shook the seats. I still remember the shock of being pushed back into the seat as we shot upward into the sky. I don't think I could breathe for minutes.

The force started letting up, and you could hear the engine quitting and starting again. That's when panic took over. I couldn't move. The whole world was flashing before my eyes. I looked over at Todd, and he was so pale, I thought he was dead. I screamed when someone touched me, discovering it was the flight attendant. She looked at me and asked if I was all right. I couldn't talk, so I only nodded and closed my eyes again. I opened them when a terrible smell assaulted my nose, and I looked over just in time to see Thomas puking his guts into a small paper bag. I closed my eyes and prayed that this was all a dream and that I would wake up in my little bed in Minot, North Dakota.

San Diego, California was about to meet some tired scared farm boys in five hours, and not one of them still felt like the king of the mountain. The eight guardians of the future were suddenly, humble, little boys who wanted to go home.

CHAPTER TWO

The plane was circling as I leaned toward the window and got the nerve to see the world from above. The ark of the plane gave me a panoramic view of the ocean, the city, and finally the landing strip. There was also a collection of yellow buildings. We seemed to be over a large airport. The pilot announced that we were about to land. We were told to put out all cigarettes and make sure our seatbelts were fastened. Actually, I think it was the stewardess that made the announcement, but I wasn't fully paying attention.

I suddenly heard a ripping noise under my seat. The whole bottom of the plane was coming loose, and I was sure it would leave a gaping hole that would suck me out of the plane. The ripping noise stopped, but what sounded like the engines cutting on and off, started again. The plane then started to weave and wobble in the air. The little pockets that took your breath away became more prevalent, and the shutters I felt at takeoff, began in earnest. The wheels bounced, and the plane made a few crow hops before settling in on the runway. I heaved an abbreviated sigh of relief and thought to myself, *we made it*, just as a sudden force tried to throw me forward in my seat. A tremendous roar was followed by a big chunk of the wing outside my window bending the wrong way. The plane made a few darts right and left before quickly smoothing out and rolling down the runway at a slow pace. I looked around, and the bloodless faces I saw were all my buddies.

We taxied right up to a gray building and a group of military green trucks. The program was simple with civilians on the left and enlistees on the right. The uniformed Marines were very businesslike. With amazing efficiency, we were loaded into long, open box, trucks. I noticed that some of the trucks were partially full, and I realized we were only part of the group they were expecting. The seats folded down from the sides and what little belongings we had were put in the middle. The Marines gave a few short orders, and the convoy started away from the tarmac.

The landscape along the road was like something I had only read about. The palm trees reached far into the sky with totally foreign-looking shrubs and grasses at their feet. The weight of the air was much heavier than I was used to, and the unmistakable odor of the salt sea water burned my nose. The Spanish-style buildings seemed to be sprawled out in a very informal fashion. The land was a mixture of sand, rock, and some type of tubular ground cover that I later learned was called ice plant.

The roads were all paved, and as we left the airport, I was

mesmerized by the amount of traffic and completely frightened. The trucks entered a six-lane highway with three of those lanes sending cars in the opposite direction, while thousands of them dodged in and out of the way. The volume and noise of the highway didn't compute in my brain. The freeway wound in and out of the city. The whole road would lift many stories high, and the circle-eight interchange above would interweave with a breathtaking view of cars below, to the left and the right. The volume and speed of the traffic reminded me of the Buck Rogers movies we had spent Saturday afternoons watching and fantasizing about. We all cowered in the truck, and the landscape became secondary to survival.

The trucks approached a long, high, cement wall that seemed to run forever. The buildings behind the wall were all the same color and neat order of the government. I sensed, rather than knew, that this was going to be our destination. The turn from the freeway into the base was sharp, and the truck jumped the curb as we came to a halt in front of a red and gold sign. It said, "You Are Entering Marine Corps Recruit Training Center, San Diego, California".

The trucks pulled over to a marked-off area, and a Marine who I guessed was an officer, walked smartly up to the first truck. The conversation was short, and the officer spun around on his heel and saluted the driver.

The first driver put his left arm out of the window and gave a pumping motion to the rest. The diesel engines puffed black smoke, and we jolted into motion. The military police at the main gate waved the trucks through, and then we proceeded along a two-lane road that led towards a huge complex with a large asphalt field in the middle. I'd never seen such a large expanse of asphalt. It seemed to be hundreds of yards across and even longer than it was wide. The truck stopped in front of a huge building with block letters, RECEIVING BARRACKS.

The trucks pulled up into a tight formation. One of the Marines came to each truck and instructed us to sit right where we were until he told us to move. The Marine told us to have a smoke if we wanted. He told us not to stand up and not to talk. He walked into the building, and we were all alone. We all started talking in a low tone, and as time passed, the volume increased. The Marine came out and said, "NO TALKING," in a serious voice. He also ordered us to put out the cigarettes and make sure not to leave the butts on his truck.

We were told to unload and line up on the stripes painted on the asphalt. Several groups of Marines could be observed all over the base.

The most spectacular array of troops was out in the middle of the huge asphalt field. They were marching in exact formation, and the bone-chilling cadence of the one marching alongside made the hair on my neck stand up.

We watched the precision platoons march in as they moved in our direction. There was a loud popping noise as hundreds of boots hit the asphalt at one time. It sounded like gunshots as they went by. The Marines made movements in unison, and we were awed by the breathtaking performance.

The Marines were dressed in green military fatigues, boots

that shined in the sun, and caps pulled down low over their eyes. The fatigue pants were tapered at the ankle to fit tight at the top of the boots. They all carried rifles set against their right shoulder.

I could observe a large number of Marines in the distance. They were dressed in green pants, yellow shirts with red letters, and white tennis shoes. They seem to be running, crawling, or moving in and along a long series of high scaffolding. I remember the huge dust cloud that hung over the area.

A marine wearing a 'Smokey the Bear' hat stood in front of us. He introduced himself as Staff Sergeant Smith, and he said we should consider him our mother until our drill instructors arrived. This struck me as funny at the time since he was black, and I laughed out loud. He was in front of me so fast it scared me. He leaned into my space with his face only inches away from mine, and started screaming at the top of his lungs that I couldn't move or talk or do anything without his permission.

The words were delivered with a generous supply of saliva, which ended up all over my face, and I made the mistake of trying to wipe it off. He really came unglued. He was so close, I could smell his lunch. The saliva started flying into my eyes, and the screams were deafening. I stood frozen in my spot. He ranted and raved and threatened all of us if we even breathed before he finished speaking.

S-Sgt. Smith told us we would be living in the building right in front of us until the platoon was officially formed and we were moved to the training area. He pointed towards some round, silver, Quonset huts in the distance and told us that it would be our final destination.

First, we had to be processed and issued uniforms and such. He repeated that he was not going to be our drill instructors, but he was stuck with us until the real ones showed up. He said he was glad it was only temporary because we all looked like a bunch of farmers and would be dead before the end of the week if he was in charge. He told us that he hadn't officially sent in the count of how many had arrived, so they wouldn't even know if he killed a couple. He looked at me when he said that, and damn if I don't almost believe him!

We were told to stand on the stripes until he came back. The Sergeant went away, and about one hour later, he returned to find most of us sitting down, smoking. Some were laying down sleeping. His screaming brought another five Marines, and they proceeded to kick our asses and take names. You could tell they had done this before. They hit us in the right places. They kicked us, so we stayed down. The total mess of recruits were back on the stripes that were painted on the asphalt in less than three minutes. The ones that were smoking had to eat their entire cigarette butt, filter and all.

The half dozen Marines kept going up and down the lines hitting and reiterating loudly that we were not allowed to do anything unless we had permission. When they thought they finally had our attention, the black Staff Sergeant decided we needed some exercise to loosen up our cramped joints from standing so long. He ordered all of us on the ground on our bellies. He said he wanted us to do push-ups. He said a proper

Marine pushup was to keep the body rigid and to only touch the fingertips and toes on the ground. The proper starting position was in the up position. He instructed all of us to get into the starting position. The six Marines then walked around stepping on hands or knees that touched the ground. There would also be a kick to your belly if it sagged.

The black Sergeant came up behind me. He crouched down and grabbed me by the hair, tilting my head back. He asked me if I was the smart ass that laughed when he said he was going to be my mother. He told me we all looked alike, so he couldn't tell. He informed me that all white people looked anemic and that it was his mission in life to rid the world of us fish bellies. I was on my toes and fingertips, so I couldn't move. Two other Marines joined him just as I said, "Yes." The sergeant kicked the side of my knees, and suddenly, I lost all feeling in my legs. I dropped to the ground and just laid there. I couldn't move my legs. One of the other Marines ordered me to stand up. He said I was a faggot and either stand up immediately, or he was going to make me suck him.

I grabbed the legs of another instructor and started to halfway pull myself up or *him down*. Another blow was delivered to my collarbone area, and I again hit the deck. Now my *legs and arms* didn't work. The one standing in front of me kept kicking me in the face, gently at first and then harder as they kept giving orders for me to stand. I couldn't. The kicks to the face ceased just as blood began to flow out of my mouth and nose.

As the feeling returned to my legs, I struggled to my feet. The black Sergeant walked in front of me and struck me in the belly so hard, that the wind was knocked out of me, and I doubled over. Another kicked me in the ribs, and I was back on the deck. I gasped for air as I listened to the Marines explain that they were "god." We would only do and say what they wanted. Anyone that talked back or disagreed would die. To make their point sink in, they each gave me a few well-aimed parting shots and left me lying there as they moved the rest of the group to another spot.

Everyone else was again put in the pushup position, and the harassment continued. This went on for so long, that almost everybody had fallen to the ground before he ever gave the command for the first pushup. He then told us we were so out of shape, that he wouldn't allow us to even do a Marine Corps pushup.

He ordered me to join the group, and I quickly stepped into a line. He told us to all stand and run in place. We were a comical bunch trying to do as he commanded while the six of them pushed and shoved us. The first few that fell on the ground got their legs tangled up with the rest. Soon, all of us were on the ground or falling, and the S-Sgt. was screaming orders. The six went about their work efficiently. I think they kicked, hit, or tripped every one of us. The blows were all calculated to inflict maximum pain with the least amount of physical evidence, in case of a serious injury.

We were taken inside the building. The room was a long concrete enclosure with double bunks on both sides. Twelve-foot-high windows filled one wall. The bunks fit in between the four-foot-wide windows with a locker next to each bunkbed. The floors were concrete and glistened with cleanliness. The stark metal bed frames and the rolled-up mattress reminded me of the Al Capone movies showing the inside of Alcatraz Prison.

The S-Sgt. told us to walk in a single file to a room where a pasty-faced Marine threw a blanket, two sheets, and a pillowcase to us and yelled, "Next!" The S-Sgt. told us to pick a bunk, make it, and then stand in front of it until he told us to move. No one knew how to make a bed. The top bunks were hard to reach, and the blankets were not cooperative. The whole thing was, to say the least, a study in chaos. The other five from the welcoming crew kept insulting and shoving us around until all of us were standing in front of our bunks. Then they disappeared as quickly as they had appeared. The S-Sgt. told us all to fall outside, stand on the stripes, and get ready to go to chow. He told us to try and stay in some sort of formation as he herded us down the huge asphalt field towards a long low building. It had hundreds of people in all forms of dress waiting in long lines that surrounded the building.

We moved very quickly into the mess hall. The first station we came to had large piles of tin trays. Each tray was divided into several sections. The next station had large round pans holding knives, forks, and spoons. The next had metal glasses, and, finally, we entered the serving line. The Marines behind each food tray had large spoons that they used to ladle out a preset amount and slop it onto our metal tray. The food line seemed to go on forever.

They just kept piling food on if you didn't move quickly. I had all sections overflowing by the time I reached the end. The line then flowed right to long wooden tables with benches.

We filed right in just as we went through the line. The condiments were all on the table, and you were expected to eat and get out fast, as the never-ending line just kept flowing. We were pretty much left alone, as we were the only ones in the mess hall with civilian clothes on. The rest of the groups had all kinds of different get-ups. Some were in shorts, some in battle dress, some in full-blown Marine uniforms. They all looked at us and silently laughed or whispered some kind of "YOU'LL BE SORRY," or "WAIT UNTIL TOMORROW" type of thing.

We noticed that the Marines in uniform all ate in time with one another as they sat at the table looking straight ahead. We all finished as much as we could of what we were given to eat, and then we were ushered out the side door. We were told to clean and scrape the metal trays into a G.I. can and put the silverware into huge vats.

We all stood in a rough formation and awaited the return of the Sergeant. The S-Sgt. herded us back to the receiving barracks and told us it would be lights out in one hour. He told us to shit, shower, and shave. He said that the next day would be a full one, so be sure to rest.

He stopped in front of me and just stood there looking at me. I asked him if they had an alarm clock. *It was funny*. He just smiled and said he would remember me. We all gathered in small groups and discussed the day's events. We had certainly experienced some new things and met some weird people, but it wasn't that tough. We laughed and told each other how we weren't scared. I told Adam and Vacky that I was going to knock that jackass of a Sergeant on his butt if he spit on me again. They all knew I was lying. Nobody was going to throw rocks in glass houses, so they let me get away with the idle boast.

Todd asked me to sneak outside with him and try to find a place to have a cigarette. We hadn't been allowed to smoke most of the day. We went down some back stairs into a small room used to store mops and such. There we sat and inhaled two smokes. Little did we know that they would be our last for some time to come.

I lay in the top bunk unable to go to sleep. My whole world had changed in the last few days, and my body and mind were revolting. I had never been put under such stress, and it wouldn't compute.

I kept hearing my dad saying, "Don't be stupid. Stay here, and go to school." I recalled the tough guy image we had in Minot. We were king of the hill. We were invincible. We could handle anything life could throw at us. At that moment, I didn't feel like the strutting young man that had defied all in North Dakota. I felt like the lost little boy who had just arrived in San Diego, California and thought the world as I knew it had ended. Somewhere in the dark room, I could hear crying. I listened as a few more chimed in. The exhaustion from my first day in the Marine Corps took over, and I sank into a comatose-like sleep that ended up being invaded by a dream of Mindy Chambers breasts.

They were so soft and so tantalizing that I reached out to touch them. Mindy looked at me and opened her shirt. Her nipples were hard and standing out. I put my fingertips on them, and she sighed. I moved my mouth towards the mountainous globes, and I could taste the woman's smell. Suddenly though, the breasts disappeared. They were replaced by the ugly face of the Sergeant.

He had one of Mindy's breasts in each hand, and Mindy was moaning as she unbuttoned his pants. A huge penis jumped into the air. The black S-Sgt. laughed as he swooped away with Mindy and told me I would never see a female again. I saw him flying into the air, and Mindy was laughing at me. After the dream ended, Mindy turned her back on me and said goodbye.

I was suddenly awakened by Adam. He was pulling on the side of my bed, crying. I asked him what the hell he was doing, but he could only cry and utter gibberish. I told him to get the hell away and tried to roll over. He let out a high moan that I was sure would bring the Sergeant and the five other antagonists, so I grabbed him and put my hand over his mouth. Todd and Vacky joined me, and we dragged Adam back to his bed. The evidence was apparent as to why he was scared. He had wet his bed.

The sheets and the blanket were on the floor, and the scent of ammonia was heavy in the air. We knew he had kidney problems, and it never dawned on us that we would be dragged into this adolescent problem of his. The lights came on, and the sergeant stood there. Adam let out a whimper, and cowered in the corner. The S-Sgt. came close and observed the evidence. He looked at all of us, and I prepared for the havoc he had planned. The Sergeant surprised all of us and asked us to take the soiled linen and blankets to the front. He said to stack them against the wall and return to our racks. He then asked Adam to accompany him. Adam became completely irrational and crawled under the next set of bunk beds. The Sergeant tried to get Adams out from under the bunks, but Adam was like an animal. He screamed and kicked and pulled the whole rack over. Their bodies were all intertwined on the floor, and the Sergeant grabbed Adam and cold-cocked him.

We watched as the guards on duty took the unconscious Adam away, and we were ordered to go to sleep when the lights were turned off. I didn't think I would close my eyes again that night. In fact, I fought to stay awake because I knew that Adam was dead. They had killed him, and it wouldn't have surprised me if they came back for more!

I must have dozed off because the next thing I remembered was the lights coming on. Day number two had started.

LETTER #1

We arrived in California. We flew right into the middle of the city. I was a little nervous on my first plane ride. The plane made a lot of grinding noises as we took off. I suppose that is normal.

The Marines picked us up in large trucks that I now know were called cattle cars. The guys who were in charge sure hollered a lot. They expected us to know exactly what they wanted.

We were taken to a really big building, where a Marine in a 'Smokey the Bear' hat told us to stand at attention. He said his name was Sgt. Smith, and he was our mother. I laughed, and he put his face right in front of mine and screamed so hard that spit hit me. Another Marine came up behind me and hit the back of my knees with a club. I dropped to the ground, and they proceeded to instruct me.

We are staying at a building called Receiving Barracks. I guess it won't be so bad. We got to go to supper at a huge mess hall. You won't believe how many Marines were in this place. Some had uniforms on, and some were in gym shorts and yellow sweatshirts. I guess they must have been doing some sort of exercise. I thought it was funny that they were wearing boots with their gym things. Maybe they were walking in a rocky area.

The Sgt. took us back to the receiving area and told us all to sleep tight. He said we would be processed tomorrow. He said we might have to get up a little early, as he had a full day planned. I asked him if we needed an alarm. He just smiled and said he would remember me. This won't be so bad!

LOVE, DALE

CHAPTER THREE

I was floating in the sky. The scene below me was a mountain valley. The air was crisp, and the first snow made the trees appear painted. The stream that ran along the bottom was bubbling, and the deer standing by the edge drinking was a perfect four-point muley. I couldn't figure out why I was here. I seemed to be floating in spirit only.

The deer wasn't scared, so I must be invisible. I reached out to touch him, but something frightened him. The sudden flash of light and the yell of another human made him disappear.

My dream was rudely interrupted by a Marine yelling, "Reveille, reveille, up and at em. Get out of those racks; hit the deck, reveille, reveille."

I sat up in bed and observed a Marine with two stripes coming toward me. He was shaking and yelling at each bunk as he went by. He hit the bunk I was in with some kind of stick and continued on. He did an about-face when he reached the end of the aisle and yelled that he was Corporal Johnston, corporal of the guard.

He said S-Sgt. Smith was busy, so he was going to take us to chow. He said we had fifteen minutes to shit, shower, shave, and dress after which, he wanted us all standing on the painted stripes outside. He said anyone late would die. He said the one they'd taken out last night was already dead.

I wanted to say, "You can't kill a man for pissing his bed," but my mouth wouldn't open. I decided I didn't need a shower. I was too scared to shit, and I was still too young to shave. I just dressed, used the urinal, and stood around talking to my friends. We were trying to hide our fear behind what we thought were stoic expressions, but it just looked like we all had a funny look on our faces while trying to appear tough.

Todd asked me if I had seen Adam. I said no, but I figured he was gone for good.

Thomas delivered the next question with as serious a face as I'd ever seen on him. "They really can't kill him for pissing his bed, can they"?

I said no, but I thought they might put him in jail. I had read stories about military brigs, and I figured Adam was already in a striped suit, breaking rocks as we spoke.

I said for all to hear, "We have to figure a way to let his folks know. Has anyone seen a telephone?"

The Corporal came back into the room and yelled for us to all get outside. "NOW!!"

Some of us weren't dressed. Some still had shaving soap on their faces, and some were still laying on their beds. He pushed, yelled, and shoved us out the door.

It was a comical sight. Some men were in their skivvies, and some had no shirts on. Some even had soap on their faces, while others were barefoot. One was naked as he tried to pull on a pair of Levi's.

The Corporal started herding us towards the chow hall. We passed groups of Marines on the way. The Marines in charge of the other groups would make their charges turn their heads away from us as they passed. The Corporal halted us at a side door and had a conversation with another Marine who had stripes all over his arm; some were on top of the sleeve, while others were on the bottom. The Corporal and the other Marine laughed quite a bit and finally gestured for the three of us in front to come to them. Vacky, Todd, and I approached them. The Corporal told us to go back and line up the others in two lines at the door. He told us to make sure all of them at least had shorts or pants on. We did as we were told. The door opened, and the two lines started filing in. Chow was fast, and the rest of the people in the chow hall all had a good laugh as they observed us.

When we all got back outside, the Corporal herded us back to the receiving Barracks. The Corporal told us to go back to our racks and strip off the sheets and pillowcases. We had to fold the blanket, return the mattress to how we found them, gather all our belongings, and fall back out on the painted stripes.

Ten minutes later, an officer with silver eagles on his collars came out to where we stood. The Corporal saluted him, and about half of us used a salute of some kind. He saluted back and introduced himself. He welcomed us to recruit training and wanted us to feel good about our future. He gave a speech about how fortunate we were to be selected to train as Marines. He wanted to know if we needed anything, and Thomas raised his hand.

The Colonel asked him what he wanted. Thomas told him we weren't allowed to smoke and that we hadn't had a cigarette since we arrived. The Colonel called the Corporal over, and we could tell he was chewing him out.

The Colonel told us all to sit down on the deck, and anybody that wanted to, could light up a smoke. He carried on with his speech, but I wasn't listening. I smoked four cigarettes as he talked. He finished and then called the Corporal to him. He gave some orders, told us all to stand up, saluted, and then left.

The Corporal instructed us to pick up all cigarette butts that were on the deck. He passed around a can to put them in, then he called Thomas to the front of the formation. He told us all to stand right where we were and not to move.

He told Thomas to pick up the can and follow him. We could hear Thomas begging, and then three more Marines came around the corner. A door slammed shut.

Thomas came running back to the formation in about ten minutes. He had a look of fear on his face, and his nose and mouth were bleeding. He stood in line for a few minutes, and then one of the Marines brought out a can of water and told him to drink it. Thomas started puking as soon as he drank it. He kept throwing up, and I looked on in horror as I realized he was vomiting cigarette butts.

The Corporal came out and told us to pick up anything that belonged to us. He seemed to take special delight in making all of us pick up the cigarette butts that Thomas had puked up and put them in our pockets. He then led the way into a huge auditorium and told us to go inside. We had to form lines, right to left, with the tallest of us on the left and shortest on the right. He had us stand at double arm's length apart and then sit on the concrete floor. The room was filling up with others in civilian clothes, and I realized that more enlistees were joining our bunch. I estimated that about 100 men were in the room when S-Sgt Smith stood up in front and told us all to stand up, get undressed, and fold our clothes in a neat pile between our legs.

Most of us stripped down to our shorts and tee shirts. The Sergeant waited until everybody was at least at that state of undress, then he asked the one closest to him to come forward. He walked around the guy and asked if he had understood the order he had given. The guy answered, "Yes."

The Sergeant went on to explain that from now on anyone who was asked to speak would start the statement with SIR and end the statement with SIR. He then asked the guy if he had understood the order.

The guy said, "Yes, Sir." The Sergeant dropped him with one blow to the collar bone. The Sergeant again repeated the instruction of sir at the beginning and at the end of each statement. He had the guy stand, and he answered, "Sir, yes, sir."

The Sergeant dropped him with a blow to the stomach. He waited for the guy to stand up again and asked him why the man had lied to him.

The guy said, "I didn't sir." The sergeant dropped him with a kick. The guy started puking on the floor, and the Sergeant kicked his arms out from under him and pushed his face into the puke. The Sergeant kept this up with about three or four others until he got tired of the game. He screamed out that when he said to get undressed, he meant everything.

"Take off your skivvies and your tee shirts. What's the matter? Are you ashamed of your little peckers?" We all took off the rest of our clothes and stood there buck-naked.

Everyone was self-conscious, and the Sergeant and his foul mouth didn't help any. He kept yelling that we were all ashamed of our small white peckers. He told us to take one hand and hold our pecker. This way our neighbors couldn't see how small they were. He threatened to cut off any that didn't hold them.

It was rather comical to me to see a hundred guys standing with their genitals in their hands.

I would guess that another ten Marines joined the Sergeant, and they started getting ready to move us around.

They came to each group, and as they called out our name, they wrote a number on our chest. They ordered us to line up single-file according to the numbers. The Sergeant screamed that he wanted us asshole to bellybutton. He wanted us tight against the man in front of us. Anybody that got a hard-on was queer, and they would kill that person. The Marine Corps didn't want any queers in their outfit.

We marched into a hallway, and a Marine handed each of us a galvanized bucket. The next one gave us a bar of soap, the next one, a toothbrush, and so on until we had all the *necessities of life*.

We were marched back into the large room, told to spread out as we had before, and then sit on the bucket.

The medical inspection was next. The Sergeant told us to stand up, bend over, grab our ankles, and stay that way.

I could tell that a group of people in white gowns came into the room. I watched between my legs as they spread the cheeks and probed the rectums of those in the rows behind me. My turn came and went, and the Sergeant ordered us to stand up and grab our peckers. He said to hold them in our hands until the Corpsman was in front of us to inspect them for drippy faucets. I watched as two of the recruits were taken out of formation ahead of me. I passed, but the guy next to me was led away.

The next check was for hernias. We were told to line up close again and stand in front of one of six stations that had a doctor sitting on a stool. We waited our turn, and the doctor would ask us to turn our head and cough as he felt the groin area. I kept watching the doctor in my row as he inspected the guys ahead of me. He seemed to be doing a lot more feeling than I remembered when I had been through the exam before.

I stepped up for my turn and tried to look straight ahead. The doctor started feeling the sides. I looked down and saw a smile on the doctor's face as he ordered me to cough again and again. He turned to the corpsman with him and asked him if he could remember a bigger one. They both laughed as I turned sideways and jerked it out of his hand. The Sergeant appeared almost instantly wanting to know what was wrong.

The doctor told him I wouldn't allow the inspection. The Sergeant took me out of the line and into a back room. Three Marines were there, and they all started hitting and kicking me. I tried to explain that the doctor was taking liberties with the inspection, but they just kept hitting me until I was on the floor. They used a few well-placed kicks and dragged me to my feet. They told me to get back in line, and if I even breathed wrong, I would be dead.

They made me get back in the same line, and as my turn came to get in front of the doctor again, two of them stood on either side of me. The doctor smiled as he and his assistant kept pulling at me and making me cough. The Sergeant suddenly realized what they were doing wasn't part of a normal inspection and shoved me ahead and gave the doctor a dirty look.

We all filed into another big room and noticed that each wall had overhead doors that were open waist high. The line started at the first door. As I reached the opening, a Marine asked me what size skivvies I wore and issued me four sets. The next opening administered socks, the next tee shirts, the one after that, yellow sweatshirts, utility trousers and a belt, cover (hat), and so on until at last two pairs of boots and a pair of white tennis shoes were given to us. We carried this all back to our galvanized bucket and were told to dress in one skivvy, one pair of socks, one pair of fatigues, one yellow sweatshirt, one cover, and a pair of white tennis shoes. The rest was to be folded as they instructed and piled on top of our bucket. The boots were to sit next to the bucket, and then we were to stand at attention.

They explained that this meant holding ourselves in a rigid stance, with our eyes straight ahead and our hands at our sides with the fingertips touching the seam in our trousers. Our heels had to be together and our toes at a forty-five-degree angle. The only thing you could do at attention was breath and that only, slightly.

The Sergeant took us outside as we finished and lined us up in four lines with the tallest of us in front and the shorter ones in the rear. When he had about seventy or eighty in a group, he had us turn sideways and stand at ease. This meant your feet were to be about eighteen inches apart and your hands behind your back. We remained this way until everyone was outside.

The sergeant walked out in front of the entire group. He said that everything in the Marine Corps was done as a team. He said that if one of us had to piss, then all of us would piss. If one of us smoked, all of us would smoke.

He said the proper command in the Marine Corps if it was time to smoke a cigarette was, "At ease, the smoking lamp is lit." This command would give everyone permission to light a cigarette and smoke until he said, "The smoking lamp is out."

You would field strip the butt at that time, roll the paper into a small ball, and deposit it where it couldn't be found.

The sergeant said, "ATTENTION! The smoking lamp is lit."

Two people reached into their pants pocket and pulled out a package of cigarettes. The sergeant quickly reached them and rendered them helpless with some well chosen blows. He went on to tell us that they had broken two rules. One, he had told them exactly what to bring out of the room and what to leave. Two, he had told them exactly what they could and couldn't do at attention. They had violated both rules, and he had to punish them.

The sergeant then said, "Platoon, at ease. The smoking lamp is lit." Three people reached for cigarettes. He quickly put them on the ground and explained the rule about bringing anything out of the room again. He said we had two minutes to go back into the room and deposit in our bucket anything we had on us that he hadn't authorized.

We made a mad scramble to return billfolds, money, cigarettes, and anything else other than the uniform he told us to wear.

We fell back into our positions and awaited the next round. The S-Sergeant stood in front of us and thanked us for our cooperation. He said that maybe he had been tough on us. To show his good faith, he was going to do us a favor. He called out, "ATTENTION." All of us stood razor sharp. He said, "good, good. Now at ease." We moved into a position with our feet apart and hands behind our back.

"Good, good. The smoking lamp is lit." We all stood there as he pulled out a cigarette, smoked it, and then field stripped the butt. He then rolled the tiny paper into a ball and stuck it into the pocket of the closest recruit.

He then gave the command, "The smoking lamp is out, "ATTENTION!" (you could feel the longing roll off of us in waves) The sergeant walked amongst us, hitting and kicking us indiscriminately as he again explained that the Marine Corps did everything together.

If he smoked a cigarette, we all smoked a cigarette. It didn't matter if we didn't have any. That was our problem, not his. We were to be always ready. We were to improvise, adapt, and overcome. We were trying to be U. S. Marines.

We must learn to obey orders. He stopped in front of me and asked me if I had smoked a cigarette. I answered smartly. "Sir, no Sir."

He walked around me and asked me what nationality I was. I reported, "Sir, German, Sir."

He then said my mother must have been a whore because I had such a big pecker. She must have slept with a black man, and I must be part black.

I wanted to hit him so badly, but something inside told me to be quiet. I realized he was just baiting me to do something stupid. He then asked me if I knew what a pimp was.

I answered, "Sir, No, Sir."

He told me that a pimp was someone who had mostly bad

white blood mixed with a little good black blood. He said my whore of a mother must have let most of the good black run down her leg and only a little bit made me.

I just kept my mouth shut and he laughed and told the rest that whenever they talked to me, they had to address me as 'The Pimp'.

The Sergeant turned us over to a Corporal who was assigned to take us to chow. He had us line up on the stripes, do a left face, and start towards the chow hall. We walked to chow, and the Corporal stopped us and had us line up behind a group of Marines in dress uniforms. He was talking to another Marine in a 'Smokey the Bear' hat, and I overheard the conversation.

The group in front was going to graduate the next day. They had a seasoned look about them, and they didn't even pay any attention to us. They were dressed in utilities just as we were, but they looked different. Everything fit them perfectly. The boots were highly polished with the tops folded under. The brass buckles flared in the sun. The hats on their heads were shaped, and the brown/tan line on one of their heads was distinct as he adjusted his hat. They had no hair.

We, in turn, looked like a total misfit outfit. The uniforms were still wrinkled and ill fitting. We were wearing yellow sweatshirts that were too big for everybody. The tennis shoes looked clean and unused. We still had long hair, and we didn't know how to stand.

I knew right then that we would all die because it was impossible to take us and develop this assortment of failures into what stood in front of us that night. The Corporal came back, and the line started to move. The chow was good, and most of us ate everything they gave us.

We filed outside, and I was surprised to see the group that had been ahead of us already standing in formation smoking a cigarette.

The corporal told us to line up and said, "At ease. The smoking lamp is lit." We all stood there and expected to see him whip out a cigarette and torture us as the sergeant had. We were surprised when he asked if anybody smoked. We were all too frightened to raise our hands.

Suddenly, he realized what was wrong. I could hear him cussing under his breath, and he reached into the top of his left sock and produced a package of Pall Mall. He said not to be afraid. He offered a cigarette to as many as he had and then used his lighter to light them.

He told us all to take a few drags and pass them on to the ones who didn't have one. We all smoked our first legal cigarette without consequence in days.

The Corporal gave us plenty of time, and we watched as the Marine in the 'Smokey the Bear' hat called the platoon next to us to attention. They were perfect. Their movements were exact, and the commands were obeyed to the letter.

Soon they were running down the asphalt field chanting some kind of strange song. They passed out of sight, and the sound kept playing in our ears. Our whole platoon felt very small in their presence.

The corporal put out the smoking lamp and herded us back to the receiving barracks. The corporal told us to file into the building, retrieve our cigarettes, and return to the painted stripes. He ordered us to sit down, light up, and ask any questions we had. We couldn't believe it. Everyone was uncomfortable as we lit up and looked at him.

The corporal started out explaining that he had just gotten back from overseas. His outfit was a special company that was sent to foreign countries where the local governments needed help against insurgents. He shared that his only job was to teach the local militia to shoot the 3.5 rocket launcher and the fifty-caliber machine gun, and to handle plastic explosives.

He was back in the states on medical, and he had been assigned temporary duty at M.C.R.D. (Marine Corps Recruit Depot). He explained he had been through bootcamp here, and although we might not think so now, we would survive.

We would live through it, and some of us would graduate as Marines. I raised my hand and asked how many of us would die here. He laughed, but then a serious look came over his face. He said that very few would die, but some had in the past, in accidents. He said the training was tough, tougher than anyone else in the world. He said we had to be tough to be Marines. We had to be the best.

The corporal went on to say that only about 30 of the 80 sitting among us would graduate. The rest would fall by the wayside and be given back to the civilians. The ones that did graduate would have to have heart. The pressure would continue until the day we graduated. He said we all would experience fear. We all would consider giving up and or escaping and that those thoughts went through every mind in M.C.R.D. He said the biggest problem would be the physical demands on our bodies. The training made demands that the body couldn't give. We would fail in some tasks, and some of us would be hurt in accidents. The attrition was high, and a few would break mentally.

The job of the drill instructors was to try to break us physically and mentally. The Marine Corps didn't want average people. He went on to say that he thought we were destined for a foreign war in the near future.

He had been to some countries in the east that were fighting to gain freedom. He named some countries I had never heard of, such as Cambodia, Thailand, Afghanistan, Viet Nam, and China.

I decided he was a stooge who had been sent to warm us up and prepare us for what was to come. He became quiet and said the smoking lamp was out.

We cleaned up the area and went inside. He came into the squad bay, told us we had better get some sleep, and that our regular drill instructors would be there in the morning.

His parting words were "GOOD LUCK, YOU WILL NEED IT"! WE were just starting to get into the beds when Adam appeared at the door.

Two Marines told him to get to his bunk. We waited until they were gone, and then we all crowded around him. He didn't look dead. He didn't even look hurt. He told us that the Sergeant took him to sick bay. The doctors checked him over and put him on some kind of medicine. He was treated well, and they returned him to us. The only thing was the corporal of the guard had to wake him up every hour to go to the john. We all laughed and welcomed him back to hell.

Todd and I snuck down to the broom closet to have a smoke. We were discussing the day's events when the door opened and S-Sgt. Smith stood there. He told us to follow him. He led us down a hall and into a small office that had a desk and a chair in it. He told us to face the wall and stand at attention.

He proceeded to sit down at the desk and do something. We stood there scared to death. The minutes dragged into hours. The sergeant never said a word. We just stood there. Todd was the first to pass out.

We didn't know enough to not lock our knees. He fell to the floor, and the sergeant just kept on with whatever he was doing. I started to feel faint and attempted to move. The Sergeant only gently nudged me, and I tightened up again.

The locked knees took me out just about the time Todd was coming around. The sergeant just told him to remove me from his office and to go back to bed.

We both went back to the squad bay, and I went to bed so exhausted, that I only remember going toward the bunk, not laying down.

LETTER #2

Well, we are starting another day. A Marine who said he was the Corporal of the Guard woke us up by slamming on G.I. can. He said Sgt. Smith was busy. He said he would take us to chow breakfast, and then we would be processed.

They had us all file into a huge room. Here, some Marines in uniform started hollering at us to get undressed. We all took off our clothes and piled them on the floor. This Marine screamed that he meant all of our clothes. Mom.

We had to take off our underwear too! We all stood around naked as a Jay Bird! There were a lot of us in this room; it seemed like hundreds. They then screamed at us to line up, single file, one right behind the next. A Marine went to the head of the line and screamed he wanted the line A to B! Mom, this means you have to stand right against the guy in front of you. I mean touching! He then ordered us to march into the next room. Here each of us was given a galvanized bucket. We then proceeded to receive a toothbrush, a shaver, a bar of soap, and a towel. They then made us all line up sideways and told us to sit down on the bucket. You have to visualize a hundred or more, naked, sitting on a bucket to really see the funny side of this!

We didn't get dressed for hours. They pushed, prodded, walked, and humiliated us from one room to another. They finally let us put on a pair of green pants, a yellow sweatshirt that said U.S. Marine Corps, and a pair of white tennis shoes.

We were taken to chow and then back to the receiving area. The Marine said our drill instructors would be here tomorrow. He smiled and said, "Good luck, you will need it!"

This won't be so bad.

LOVE, DALE

CHAPTER FOUR

Shrouded in total darkness, my mind could not fathom the noise and the bright lights that were thrust upon it. The only thing that registered was hell! I had died, and I was at the gates of hell. The devil was roaring up and down the walls of a canyon, and I was falling deeper and deeper into the pit of hell.

I suddenly realized that I was not dead. The noise was a man in a 'Smokey the Bear' hat, and the roaring was a metal garbage can as it rolled and banged along the concrete floor. The man was screaming, "Get up you maggots," and kicking the can when it stopped. He moved through the barracks and turned at the end.

During his next trip, he started pushing the double bunks and pulling the covers off the frightened recruits. He shoved the bunk I was on, and I grabbed on with both hands. He screamed, "Up maggot," and pushed the rack into the middle of the floor. I hung on for dear life and noted the recruit below me was doing the same. The man with the 'Smokey the Bear' hat hooked one foot on one of the legs, pushed just right, and the bunk flipped over. I was suspended in midair for a heartbeat, then dropped like a rock onto the concrete floor.

The man screamed, "Attention, you bastard, attention!" I scrambled to my feet and stood next to the recruit who had been on the bottom bunk. The man stood right in my face. He was only an inch from my nose as he screamed, "You maggots will move when I say move." His saliva was hitting my face, and, suddenly, he kicked my leg. I dropped to the floor and rolled as he followed up with a foot to my stomach. He moved down through the barracks dealing out insults, kicks, and hits until he reached the other end.

The man in the 'Smokey the Bear' hat yelled "Attention! Stand at the end of your bunks." We all tried to become invisible as we waited for the next round. He told us that his name was Sergeant Rokel. He was one of our permanent drill instructors, and we belonged to him for the next twelve weeks.

Sergeant Rokel looked at his watch and began yelling that we were late. It was already 03:20, which left us only five minutes to shit, shower, and shave.

"You maggots will be standing out on the painted stripes in five minutes." The Sergeant kicked, shoved, hit, and threatened us for the next twenty minutes as he tried to get everyone outside. He stepped in front of me and told me to have everyone line up to height. I didn't understand what he meant, and I was dumb enough to ask.

He spit in my face and screamed that he wanted the tall men in front and the short men in back. I scrambled and tried the best I could, but I was only met with more abuse. I finally accomplished what seemed to appease him, and he shouted, "Attention. Right face. At a run. Forward march."

We turned to the right and understanding that he wanted us to run, the ragged bunch of us lurched and seemed to run as a herd.

The D.I. moved us out to the asphalt area and screamed, "Halt." We all skidded to a stop with our formation looking like a gaggle of geese at feeding time. He pushed and kicked us into a facsimile of a formation and called attention.

Sergeant Rokel pointed to the area of the chow hall and said we were to run in formation to breakfast. We were not to break formation or stop. He then told us that the proper command for running in formation was quick time march. He shouted, "Right face. Quick time. March!"

We set off at a slow pace in somewhat of a formation for the chow hall.

I guess we'd run about a half mile before the first of the recruits dropped out. The Sergeant yelled, prodded, and pushed the fallen recruits back into the formation. He kept this up until all of us were on the ground in some form of exhaustion. I threw up right in front of the sergeant. Surprisingly, he just kicked me once and moved on.

We slowly made it to chow. The sergeant lined us up at the door and informed us that we had five minutes to enter, eat, and be back there in formation!

The chow line moved very fast. The messmen that served chow ladled the food all together. The culinary delights for that early morning's meal included scrambled eggs, (so tough that the fork couldn't cut them) toast, (kept in a steamer that made it wet as a sponge), hashbrowns, (good), pancakes, (cold), chipped beef and gravy poured over everything, and your choice of milk, juice, or water.

We filed onto benches that ran along the sides of the long tables that seemed to run forever. We wolfed down as much as possible and filed to the door that led to our formation area. The D.I. and three or four cooks were waiting for us. The cooks told us that the Marine Corps was on a budget. The food we took was a balanced diet, and it was all we needed to be healthy.

Still, a rule in the Marine Corps was, "Take what you want, and eat all you take," which meant we had to clean everything left on our trays. The D.I. said we could not use our utensils. We had to lick it up with our tongues. It was quite a sight watching 80 or more people trying to lick up the slop left over. The D.I. pushed us into formation as we finished. When the last of our motley crew was ready, he issued the following orders, "Attention, right face, forward march," and we marched towards the receiving Barracks. Most of us had eaten much more than needed, and the walk felt good. We were surprised to hear the following order, "Platoon, QuickTime, March."

We broke into that half run the Marine Corps is famous for, and the full bellies started acting up right away. The first few that started throwing up were left behind. The main body of recruits all started feeling nauseous at about the same time.

The D.I. made us run in a circle until the next set of recruits finished upchucking and rejoined us. This continued until all had their turn on the ground. The D.I. screamed, kicked, and pushed us until we finally arrived at the receiving barracks.

The D.I. left us standing at attention on the painted stripes we had come to know so well. We must have waited for two or three hours. The pain of forcing yourself to remain immobile cannot be described.

Thank God I had learned earlier not to lock my knees. It was weird to see the guy standing right next to you just kind of jump forward and fall without moving their arms. They would hit the ground and bounce. They usually fell on their face, and the little puddle of blood from their nose was the only thing moving.

The D.I. returned with two more Marines. The three of them wearing 'Smokey the Bear' hats were very intimidating. They moved to the front of the formation, and Sergeant Rokel said, "At ease." Most of us didn't remember what it meant, and everyone did something different.

He screamed, "Attention!" as we all stood frozen in fear. Sergeant Rokel went on to explain that "At ease," meant to put your left foot approximately 18 inches out and put your hands behind your back, placing the left under the right in the small of your back, palms outward. He issued the order again, and most did what he wanted. He went on to introduce the other two men.

"This is Corporal Val. He is one of your drill instructors. Corporal Val and myself will be your daily contacts with the Marine Corps." He motioned to the other 'Smokey the Bear' hat wearer. "This is Staff Sergeant Rowel. He is your head drill instructor. He will be in charge of everything in your life from this moment on. You now belong to the three of us. You will not talk to anyone else, and everything you need, will be provided by us." We didn't know whether to be terrified or just weary.

"I want to share something about all of us with you." He gestured to the man next to him.

"Corporal Val was brought here from Force Recon, a unit of the Marine Corps that trains killers. Corporal Val has one problem." He sighed. "We are not at war, so he can't get the one item he is missing to become a full-fledged member of Force Recon. He must kill three humans and bring their ears to the commanding officer to be a full member. Corporal Val volunteered to train recruits in hopes that at least three of you would give him an excuse to kill you, so he can fulfill his obligation to Force Recon."

No one moved. I don't think anyone breathed.

"I am Sergeant Rokel, as you know. I am a drill instructor because the Marine Corps would not allow me to be in a regular unit. I was shot in the head, and sometimes I black out and don't remember what I am doing or where I am. I have a habit of shooting people when I get this way, so they don't allow me to have a weapon. The worst thing that can happen here is I might kill a few of you, and the Marine Corps won't care about you, as you will learn in the next few weeks. Some of you will die anyway, so how it happens is in material."

He leaned his head to his other side to indicate the man on his other side. "This is Sergeant Rowel. He has graduated seven classes of recruits at this Depot. You will be his eighth and final class. Sergeant Rowel is here because he is too old for the regular corps. He was shot in Korea. He only has one lung. The Marine Corps has told him that he must retire after he is done with you, but he can stay in the Marine Corps if he can make all of you fail.

"This is what you are up against. We will tell you when to get up, when to go to bed, when to shit, when to eat, when to talk, run, walk, and crawl. Though you will depend on us to keep you alive, we can kill you whenever we want.

Nobody can help you. We are the only chance you have to live. You do what we tell you, and some of you might get out of here. Last week on Paris Island, a drill instructor took his whole platoon out into the swamp and drowned some. We will do the same to you if you even hesitate when we give you an order.

The Marine Corps is built on teamwork. You will be trained as a team. If one of you messes up, all of you will pay for it. You will learn to discipline your peers. If one of you can't get it right, we expect you to take care of him as a team. You have our permission to beat any malingerer to within inches of his life. Only the three of us can kill. I am going to turn this over to Staff Sergeant Rowel and see if he has some words of wisdom for you."

"Good morning recruits." We all stood silent. Staff Sergeant Rowel said, "I just said good morning, and you didn't acknowledge? Let's try that again!—Good morning Recruits." We let out a weak reply.

S-Sgt. Rowel replied, "I don't like that. All of you get in the push-up position." We all dropped to the ready position. The Staff Sergeant smiled and started walking amongst us.

He said, "I see you already know the ready position. Good, now count with me. Down and back up is one, down and back up is two and so on. Ready. Begin, one, two, three, four — Wait. Some of you are doing it wrong, start over."

"Sergeant Rokel, Corporal Val, help me. This bunch doesn't seem to understand directions." His voice was directed away from us.

"Start over, one, two,— No, start over. One. Start over, one, two,— Sergeant Rowel, help that one there."

It went on like this as the three of them kicked and stomped on hands until all of us had dropped to the ground. Staff Sergeant Rowel called us all pussies and ridiculed us because we never even got to six push-ups. In reality, we did 50 or more, as well as remained in the ready position until we fell. He turned it over to Cpl. Val after that.

Corporal Val called us all to attention. He had us line up on the stripes and then gave the "at ease" command. We all did it right, and the Corporal said, "Excellent." He then rambled on about the honor of being in the Marines and the fact that some of us would be heroes. Some of us would be dead, and some of us would break. Others would become men.

He then said, "I need three volunteers. I want three of you to agree to let me kill you now. I do not want to be here. I want to return to my Force Recon unit. I need three sets of ears to take back. You can save your buddies a lot of grief in the next twelve weeks by letting me kill three of you right now."

He then gave the command, "Attention," as he started down the first row. Corporal Val was short. He stood about five-foot-eight or so. I was the first one he came to.

He looked up at me, and, suddenly, his hand shot out toward my throat. He pushed his index finger into the soft spot at the base of my neck and slipped under the breastbone. The pain was overwhelming. I immediately dropped to my knees, and he pushed my body backward so that my weight was balancing on his finger. He told me I was too tall. He hated tall people, and I was the one he planned to kill. He let go and continued down the row.

S-Sgt. Rowel came up to me and asked why I was not "at attention." He punched me in the ribs and ordered me up. I staggered to attention just in time for Cpl. Val to arrive and deliver a quick chop to the neck. There was another swipe to the knees to put me down again because I was too tall.

The three of them kept this harassment up until they tired of the game. We were then called to attention, moved to an area behind the barracks, and given the command to sit down in place. Sergeant Rokel gave us a lesson on Marine Corps words. He explained that we must talk in the right terminologies. The toilet was *head*, food was *chow*, a hat was *cover*, the bed was *rack*, the floor was the *deck*, and the wall was *bulkhead*. He went on with more and more examples and then suddenly came to attention and saluted another Marine who was dressed in a khaki uniform with a brimmed hat., (now known as a cover), with two silver bars on his shoulders.

This turned out to be Captain Sikla, our commander, and he welcomed us to the Marine Corps. I didn't pay much attention to the speech since I was intent on watching how nervous the D.I.s were in the presence of an officer. I figured something didn't jive. What were they afraid of? Why were they so nice in the presence of an officer? The officer finished his speech, saluted the D.I., and walked away. The D.I. waited until he was out of sight and then gave the order for us to stand at attention.

Corporal Val moved us back to the interior of the asphalt slab, put us back on the painted stripes, and said, "At ease." We stood there awaiting our fate.

Staff Sergeant Rowel came to stand in front of the platoon. He gave the order to fall out. He said we had five minutes to go to the head and be back in formation. We all made a mad dash for the toilet. I had been wiggling for at least the last hour. The bliss of relief was the highlight of the morning.

We all returned, and Sergeant Rokel headed us out to lunch. We marched about 2/3 of the way, and he then gave the order for the "quick time march." The little run to Chow wasn't bad, and we were all careful about how much we allowed the messmen to put on our trays. They still took joy in mixing everything up. They dumped the gravy on the bread and strawberries on the potatoes, but we all made it through without a beating.

The sergeant was waiting for us when we were done and lined up. The formation started back towards the receiving barracks. He marched us back, so everyone got to keep their food down. The afternoon was used up listening to lectures and preparing our civilian clothes for shipment.

The mail clerk had us all go into the barracks and bring out everything we had that wasn't military issue. We were told how to fold everything up, package it, and label it for shipment home. The sergeant told us to file by and pick up a pencil and two sheets of paper.

We were instructed to write a letter to our parents telling them how much we enjoyed being there and that it was just like summer camp, except more fun. He said we were to stand at attention when we finished the letter, so they could proofread it.

Cpl. Val came up to read mine. I handed it to him, and he just stood there for a moment. Then he shoved his fingers under my breastbone, which made me go to my knees. He just smiled as he read my letter and kept his finger in the soft part of my throat.

He asked me where I was from.

I replied, "Sir, Minot, North Dakota, sir."

He finished the letter just as the Black sergeant who had first brought us to the receiving barracks walked up. He started talking to Corporal Val and pointed at me. Corporal Val came back and walked up to my right side. I was still kneeling. He bent over and whispered in my ear, "Sergeant just told me that your mother is a whore that does Black men and gives birth to Black babies. Is this true asshole?"

I started to answer, but he gave me a chop on the collarbone and said, "He told me to watch you. He said you are a bad one and that you are going to try to kill me. Is that true, asshole?"

I just sat there and held my temper. Corporal Val started laughing and told the whole platoon to look at me. He told them that I was going to be the first to die. He hated me, and my ears were going to be his first set.

He then did something strange. He became almost friendly and started telling stories of his experiences in the Marine Corps. He told us about the training he went through to be in Force Recon. He shared the experience of being dumped on a remote island and eating snakes and bugs for weeks while he waited for a pick-up. He talked about going to Vietnam and teaching the locals how to fight. There was one trip where they rigged explosives completely around a Viet Cong general and then tied the detonator cord to his pet cat. He laughed as he described the look on the general's face as they threatened to let a dog in. The general started talking! He said they took him out, and a chopper picked him up!

S-Sgt. Rowl talked to us next. He told us all to sit down on the asphalt. We could spread out and relax. He wanted to know our statistics. We had to raise our hand if we fit his question. He asked how many of us were white, Mexican, or Black. The questions went on and on.

As quickly as the day started, it ended. The D.I.s all called us to attention. Sergeant Rokel told us to file into the barracks and prepare for "lights out." He said on the next day, we would move to our permanent barracks, and then life would start.

He told us that the whole bunch of us were nothing but a spoiled bunch of farmers, and we needed their guidance. They felt it was their destiny to break us and ship us back to North Dakota in Pine boxes. We would not live to *do* the family sheep again!

We all gathered in the shower and talked about our day. *We had to find a way out of here.* The general feeling was that a group of escaped lunatics had control of us, and the real Marine Corps wasn't aware of our existence. We had to find the *real* people.

We wanted a smoke so bad it hurt. No one had a butt. I lay down on the rack, and sleep slowly overtook my fears. The last thing I remember was a gargling noise in the shower room.

***

The room erupted into bright lights, and the sound of yelling and iron scraping against walls was coming from the shower room. I sat up in bed and looked at a medical doctor who was shouting orders. That whole end of the barracks was full of strange people in strange uniforms, and I watched as they wheeled out the hospital gurney with a body on it. Word quickly spread that one of the recruits had hung himself from the overhead pipes in the shower.

The sergeant of the guard went from bunk to bunk asking each of us if we heard anything. I remember seeing the guy. He was one of the recruits from Detroit and had joined our outfit late. He had only been there overnight. The stress must've been too much.

The first victim of Platoon 2002 was dead. We didn't know it, but he was the first of a list of casualties that would haunt us the next twelve weeks.

LETTER #3

```
Something is wrong! I need your help. The Marine
overturned the garbage can in our sleeping area
at 3 AM. He came through screaming "up and at em
Maggots," He grabbed my mattress and pulled it
off the bed. I fell right on the concrete floor.
He then made me stand at attention and spit in
my face as he yelled at me. Why doesn't he like
me? I can't seem to do anything right.

We all had to go outside and line up in a group.
He kept hollering about our formation and doing
it right. We didn't know right from wrong. I
tried to explain that, and he spit in my face
again.

He made us all run to the kitchen area. It was a
really long way! Nobody could run that far. He
```

kept kicking us if we slowed down. We got to the eating place and

He said we had five minutes to go through the line, eat, and be back in formation. This is impossible! It took more than five minutes just to get our food. I don't understand.

The people at the end of the line would tell the Marine if we didn't eat everything on our tin tray. They would put the gravy on the desert, fruit on the potatoes, and the peas in our milk. They made us stand and lick our trays if we didn't finish. It was horrible!

We all had to run back to the receiving area. It was impossible. Everyone kept throwing up. The Marine would holler at you for throwing up. Nobody had any food left in them by the time we got back.

The Marine said that tomorrow we would move to our own area. He said that they would start to make life miserable. He said we were spoiled farm kids who needed a guiding hand. He also said we should write home and tell you how we enjoyed our new life. He said to tell you it was just like summer camp, except more fun.

CHAPTER FIVE

The drill instructor sounded reveille at 03:00. The process of kicking the garbage can down the length of the barracks wasn't as alarming as it had been the day before. Every person was up and alert instantly. The memory of the body being moved was paramount in our minds.

The sergeant had us all fall in at attention at the end of the bunks. He began describing the condition of the body. He laughed and asked if anyone knew what happened when someone hung themselves.

No one answered, so he proceeded to tell us. "It seems that the body involuntarily relaxes the muscles that control your urine and bowel movements. The neck is pulled an extraordinarily long and grotesque length, and the tongue is usually bitten and out."

The sergeant went into graphic detail of each body movement, then ordered three recruits closest to the shower to go in and clean up the soiled floor.

The three were soon back out, and one was throwing up in the G.I. can that had just recently served as an alarm clock. The sergeant kicked him back into the shower and ordered me and two others to go in also.

The shower is a tiled room with about eight to ten shower heads on each side and a series of drains down the center. The room is divided by a wall that sits at a 90° angle. This keeps the spray from hitting the area that is assigned for shaving and such.

The sight of blood was the first indication of the night's tragedy. The trail started in the hallway leading outside and faded as you reached the partition that led into the shower. The next thing you encountered was the smell. The only way to describe it was that it smelled like death. The strange mixture of excrement, urine, and body odor was a smell that, once experienced, would never be mistaken for anything else. It is like the first time you hear a rattlesnake. It is a sound you know for the rest of your life.

The floor was covered with body waste, as well as the wall closest to the pipe in the ceiling. The recruit who was throwing up outside started again, and another turned into the corner to join him. I quickly reached up and turned on every shower head in the place. The smell abated as the water did its saintly job of washing away the debris.

The other recruits soon started helping, and the shower took on the normal appearance of government property. We exited a few minutes later, and the sergeant started again on the description of the body and said more of us would join the dead recruit. Corporal Val walked in at that moment, and the statement he made was chilling. "I want all of you maggots to know I only need two more sets of ears."

Corporal Val told me we had five minutes to shit, shower, shave, and be outside in formation. He said anyone late would be the second set of ears.

The run to chow was the same as the previous day. We ran in circles on the asphalt lot. The first recruit started to fall and throw up. The rest joined in, but there was a certain number that kept running. They threw up because they were too afraid to stop.

We arrived at the chow hall and milled round like a bunch of wild animals on their way to slaughter. The look in our eyes and stance gave it away to anyone that observed, that we were just milliseconds away from reverting back to caveman instincts. The order to go in was given, and we all plodded in, moving in a brainless way that must've been similar to those walking to their deaths in previous wars. We had no thoughts, no motivation, no pride in our movement or appearance, and no ability to react to any more horror!

The meal was quick, and the usual harassment continued for those who took too much. We all got into a formation and returned to the receiving barracks. Corporal Val gave the "at ease" order and left us standing on our painted stripes.

The sun was just starting to peek over the mustard-yellow buildings as Sergeant Rokel moved to the front of the formation and shouted, "Attention!" He said we would move out that day and head for our permanent barracks.

The recruitment area (in 1960) consisted of basically Quonset huts to house the Recruits, a building, behind the Quonsets that housed the heads (restroom), and the outdoor wash racks where the recruits did their own laundry. You were issued a galvanized pail and bars of soap. The racks were long concrete wash racks with overhead water spigots. We washed and hung the clothes out to dry.

We had to post guards. It was a form of harassment used to train you in Guard Duty, which would come into play for the rest of our tour!

The DI told us it was to protect our clothes from the Navy (who had boot camp across the water from us.) He said they liked to sneak over and take the Marines Skivvies!

Our trip from the receiving barracks to the Quonsets was not that far, but the D.I.s kept us going in circles for miles, so the quarter mile turned into miles! The same thing happened on the runs to chow.

The quarter mile or so was sometimes miles in circles. The marching area was a huge asphalt lot. I don't remember the distance in miles, but after marching all day, it was 100 miles by 100 miles on your feet!

The rest of the day would involve gathering our gear and some lectures from the chaplain, the Base Commander, and the Officer of the day. First though, we were going to do some light exercises.

The sergeant moved us out on the asphalt field, about 100 feet away from the building. We were told that the asphalt field was called the "grinder." It served as a training ground for marching and exercising. We were told to spread out at double arm's length with each row about three body lengths deep from each other. The first exercise was the side straddle hop. The sergeant demonstrated and then counted as we got into the act. We did about 50 or so and then were ordered to stop.

The next command was for us to assume the position for a proper Marine Corps pushup. We were joined by the other D.I.s, and the pain of remaining in the upright position began. The first ten went very well.

The first recruit dropped to the asphalt at about number eleven. The three drill instructors immediately surrounded him and left the rest of us in the up position. They were kicking the unfortunate slob, and soon more started. We kept dropping long before we reached the chow hall.

The D.I.s went about their work with glee. They kicked and pushed the fallen ones as fast as possible. Soon everyone was on the deck. The commands to perform were mixed with screams of contempt. The hopes of all were dashed when we were told to get on all fours. The D.I.s screamed that we were a bunch of sheep from North Dakota, so we would be treated as sheep. Sergeant Rokel came up behind me and bent down and asked me if I knew who my father was. I started to answer, but he kicked my arms out from under me. I fell on the deck, splattering my nose. The blood was flowing and somehow got on his boot.

The sergeant called the other two D.I.s over and showed them how I had desecrated his boots. He theorized and told the other D.I.s that maybe I should be licking the blood off his boots, or I would die right there.

I panicked and rolled onto my knees in a fighting crouch. The three circled me. The first kick caught me in the ribs, and the other dozen were directed at various locations. I was left lying on the deck as they continued to gloat to the rest.

Sergeant Rowel said, "You are all sheep. This maggot on the ground didn't have a father. His mother did sheep. He is half sheep. You all are sheep. You will not be allowed to walk anymore. You will all crawl on your hands and knees and go BAA-BAA-BAA. I don't want any words from any of you. You are a sheep."

The sergeant then ordered everyone to squat down into a squat jump position. He explained that this meant squatting, with one leg in front and the other under our ass. Our arms were to remain above our heads, and on command, we were to hop up, exchange legs, and go back into a squat.

"This is a proper Marine Corps squat jump position." He informed us that they were good for sheep. It started hurting at about ten or fifteen. The pain became unbearable at about twenty.

Most of the group was falling over and being kicked back up before we reached thirty. I cannot remember anything more painful in my life. The D.I.s took real pleasure in continuing until everyone was on the deck. The kicks just bounced off of us because the real pain was in trying to move.

The command was finally given to stand at attention. It was impossible to straighten our legs, and most of us remained on all fours. The D. I.'s were delighted. We really were sheep. We thought attention meant all fours. They gave the order to march, and we all crawled toward the barracks. Sergeant Rokel stopped us and reminded us that we were to go BAA-BAA-BAA! We crawled to the painted stripes going, "BAA-BAA!"

We were given the order to stand up and run in place, and our muscles soon relaxed. The order to use the head was given, and we all went inside the barracks.

We were reassembled outside on the grinder and told to sit. The Chaplain spoke to us first. He welcomed us and said a prayer. For the next few hours, we sat there listening to speeches from various officers. The atmosphere was friendly, and the thought of telling someone that the three D.I.'s were trying to kill us surfaced in most of our minds. One look from the three was enough to make the thought retreat into the recesses from wince it came. The trip to noon chow was uneventful as the Chaplain accompanied us. This time we walked to the chow, were allowed to eat, and walk back. The lectures were over at about 14:00 (2 P.M.), and once again, our lives were left in the hands of the D.Is.

Corporal Val called us to attention and told us to go into the barracks and pick up all of our belongings. He said we would move to our new barracks. He reminded us that we were sheep. We would not be allowed to get up. We had to crawl and BAA-

BAA-BAA as we retrieved our gear from the barracks.

It was quite a scene to see recruits crawling back-and-forth with their belongings. We couldn't use our hands because we were sheep. We made numerous trips with the stuff in our mouths. We received well-placed kicks if we forgot to go "BA-BAA-BAA!"

Sergeant Rowel took over when we had the gear outside. He told half of us to get up and go clean the barracks. The other half were told to stand up and go clean the head and the shower. I looked and noticed that some of the recruits had worn holes in the knees of their trousers, and the dark red of the fabric showed the presence of wounds to the legs from crawling. He looked at my face and told me to go in and wash the blood from my nose. He wanted me right back out to guard everyone's gear.

The formation grew as the chores were finished. The entire platoon was called to attention. Sergeant Rowel said a third of us would remain on all fours, and the rest would be allowed to walk upright and carry the gear. The third would crawl and go by BAA, BAA.

We set out across the grinder with the front third of us crawling and the rear carrying our few meager possessions. The asphalt quickly tore the trousers at their knees, and the recruits started dropping from the pain.

Sergeant Rowel stopped about a third of the way across the grinder and had us change positions. The next third dropped the equipment and got on all fours. Those of us who had been on our hands and knees got up and carried the equipment.

The second half of the journey was as gruesome as the first half, and we repeated the change one more time. We reached the other side just as one recruit suddenly jumped up and tried running for the safety of a closed building. The D.I.s caught him and pushed him into the hallway. The grunts and yells were muffled, but the recruit and one of the D.I.s didn't come out.

Sergeant Rowel and Corporal Val quickly returned and had us turn left. We continued along a dirt path that was a welcome relief to the crawlers, as the dirt covered up the cuts and cushioned the knees.

LETTER #4

> I can't believe you let me do something this stupid! I don't belong here. These people are crazy. This is the worst experience of my life. The D.I. won't listen to anything I have to say. He thinks we are all animals.
>
> As a matter of fact, he calls us a herd of North Dakota sheep. He says we don't know how to do anything. He makes us all get down on all fours and crawl. He says we must have been raised by sheep in North Dakota. He asked me if my father had wool.
>
> We went to the Quonset huts that will be our home for the next months. It is the same type of building that we store wheat in, in North Dakota. I can't believe we have to live in them. Why

don't you see if our congressman can save me! These people plan on killing me, HONEST!

We have to get up really early in the morning then run to chow and back. I still throw up. Most of us do. We have to go out on this giant asphalt field and do exercises forever, then the D.I.'s inspect our Quonset huts. We never pass! He makes us do squat jumps and then clean it all over again. Yesterday we cleaned it four times. He finally said we were hopeless and made us do more squat jumps.

I have never done something so painful in my life! You have to squat down, one leg in front of the other. Your arms are over your head, and when he gives the command you hop up and exchange legs. Then you reverse this! It starts to hurt right away. He keeps on until you fall. He then kicks you until you get up again.

This goes on until everyone is on the deck. He then makes us crawl like sheep and go BA-BA!

I don't think it's legal what they are doing to us. Someone said lots of guys die here!

It's not so bad.

LOVE, DALE

We stopped at the edge of an asphalt street. Corporal Val gave the command to stand up, and we got the first view of our new homes. The asphalt street ran between a group of Quonset huts. The area between the huts was covered with a strange ground cover we had only observed on the way to the Marine Corps Recruit Training Depot. It was called an ice plant, and they used it like we use grass in North Dakota. The huts were silver colored, and the signs were painted in Marine Corps red and yellow. The whole area had a dusty look, and it was obvious we were the first tenants for a while.

Corporal Val told us to line up in front of the Quonset as he called out our names. We then were told to enter and stand in the middle of the room at attention. He then came into each hut and assigned us a bunk by name. We were then told to put our gear on the assigned bunk, stand at the foot of the bunk, and await orders."

I was assigned a bottom bunk. I put my gear on the steel springs and stood in position at the end of the bunk. This gave me my first chance to look around. The hut had curved walls because of the semi-circle construction. The double bunks were stationed on each side with a locker between them. The center had a row of wood stanchions that were to be used later as rifle racks. The bunks were two stories high with steel springs and a mattress rolled up on the forward end. There were two wooden locker boxes at the foot. The hut had glaring electric lights on the ceiling and concrete floors that you could tell had been highly cleaned over the years.

Corporal Val announced that we were to line up in front of the center hut. This was the D.I. hut. The door opened, and Sergeant Rokel pushed a pillowcase, mattress cover, two sheets, and a blanket in your hands as each of us passed the door. We were then ordered to return to our bunks and stand at attention until given further orders.

Sergeant Rowel walked into the hut that I was assigned to and went to the first bunk. He took the mattress cover and installed it. He then took both sheets and installed them. The first blanket was placed on top. The blanket was stretched tight with the sheets folded at a precise length and angle, and the bunk was exact in every aspect. The corners were a military angle, the tightness tested by bouncing a quarter on the blanket and having a bounce back into his hand. Sergeant Rowel then told us all to make our bunks exactly as he had shown us. He ripped the example apart and said we had three minutes.

The resulting mess of bunks looked nothing like he had created. He tore the bunks apart, and we continued to remake our bunks for hours. This was accompanied by a good amount of squats and squat jumps. The sergeant finally got tired of our inability to properly make a bunk and told us to fall out in the street.

We were put in the squat jump position and proceeded to do squat jumps until everyone was on the deck moaning and awaiting further punishment. The D.I.'s called us into "four-legged sheep" attention. We were told to practice crawling in and out of our huts and that our timing was critical. We all had to be able to enter and/or leave in a preselected timeframe. That select amount of time was never disclosed. We just practiced crawling in and out of the hut for hours.

The final straw was when they allowed us to stand and instructed us to enter and exit so fast, that each time the slowest maggot got trampled. They wanted to see at least one maggot lying on the ground and bleeding each time we entered or left the hut.

The pain of the exercise and the verbal abuse intensified each time we didn't knock somebody down. The hours led to animal instinct that caused us to hurt each other without remorse or regret. As long as it wasn't us, it was okay. It was every man for himself.

Corporal Val called us to attention. He said it was chow time. We were to stand at attention. He then proceeded to walk down each row of recruits, and without warning, he hit each of us in the stomach as hard as he could. The results of this event left everyone on the deck gasping for air. He laughed and told us that we would learn to love him. We would look forward to him hitting us. We would need his guidance and the pain he inflicted. The day would come when we wouldn't feel the pain, but feel slighted, if he didn't hit us. He said he planned to hit each one of us every morning we were here. The only way out was death.

Corporal Val made us change our trousers. He wanted everyone who was still bleeding at the knees to take toilet paper and wrap their knees. He said he didn't want to spoil the ambiance for the others as they dined!

The trip to chow was at a walk. The line moved quickly, and most of us had learned not to take too much. We fell outside and waited for the trip back to our new home. Corporal Val arrived, and the walk back to the huts seemed to be normal. It wasn't until he ordered a left turn instead of a right that we became suspicious. The huts grew smaller in the background as we went out into a large field covered with cacti, sand, and darkness.

Corporal Val called us to a halt, deep in the open field. He ordered us to form two lines with the tallest to the left and shortest to the right. He had both lines face each other. The person in front of you was your partner. The next order took it all by surprise. He said the people in the lines were to wrestle each other. He would judge the winner!

This went on until he got tired of it. He had us reform into some semblance of a platoon and run in circles in the sand. We were then ordered to get in the squat jump position. We did squat jumps until everyone was on the sand.

Then, as suddenly as it started, it stopped. He said, "Sit down in place." We all sat down, and he went into reciting some stories of what it was like in the shithole countries he had been to.

Cpl. Val said he wasn't trying to be mean to us. He said we would need to be strong if we ended up in these strange-sounding places-because they lived a different life than we did. They were like animals, and if we were captured, they would torture us far beyond how the D.I.s would in bootcamp. He said it was okay to hate him; that hate would be useful. Cpl. Val told us it was hard to kill someone if you didn't have hate for them. That was the difference between the Marines and the other services. The Marines are called on to be the first into battle. You couldn't have fear. You need hate to survive. He said another Marine strongpoint was that we were trained as a team. "You must be willing to give your life for the team. That is the driving factor that has allowed the Marines to never lose a battle." He then ordered us back into a formation, and we went back to the huts!

We were ordered to clean our huts and prepare for an inspection. The inspection lasted for hours. We never did pass. The Corporal found more things than we could imagine to redo. He finally gave up and allowed us to go to the head. The balance of the evening was spent doing squat jumps.

The kid from Arkansas passed out, and nothing could revive him. The corporal called medics and told them the kid had fallen and bumped his head. The corporal called the medics and told them that the kid from Arkansas had slipped on the ice plant and struck his head on the pavement. The medic asked the corporal if any of the rest of us had bruises and cuts. The corporal laughed and said we were a bunch of clumsy farmers from North Dakota, and we hadn't gotten our sea legs yet. The medic just laughed and told the corporal to be more careful about leaving telltale bruises where the officers could see them. The medics hauled the kid from Arkansas away, and we never saw him again.

The time finally arrived for lights out and some much-needed sleep. The two sergeants left us in the care of Corporal Val who ordered us into the huts. He told us all to strip down to our skivvies, stand at the end of the bunks, and await his entrance. He said the moment we saw him enter, we were to yell out, "ATTENTION." Everyone was to assume the position with their eyes straight ahead, shoulders back, arms at the side, and not move anything, including our eyeballs! The first few times we didn't yell loud enough, or someone moved, or some other reason, because we kept doing it over and over, along with a generous supply of squat jumps. It was very late before the corporal decided he had enough fun.

The last command he issued was for everyone to drop their skivvies and stand at attention. He was going to check for erections, and if he found any, he would personally kill them. He went into a long lecture about the sickness of homosexuals. He told us how the other services must rid themselves of them, and he said some had infiltrated the beloved Marine Corps. They would weaken the Marine Corps. The Corps was founded by men for men, and he would not have any homosexuals in his outfit. Thankfully, he found none!

The order of lights out was issued. We all were told to be in the bunk in one minute, no talking, and to sleep at the "attention" position. The Corporal said he would come through and inspect at various hours, and if he found anyone not sleeping at attention, he would make all of us get up and do squat jumps.

The quiet was blissful. The entire area seemed to be at peace. The entire hut full of recruits were laying in bunks at attention. The sounds of sniffling were muffled as they were interrupted by the perfect quiet and whispers between bunkmates that sounded like the winds roaring through the Pines in Montana.

I knew the D.I. was going to return, and I would have to pay for the few weak links. I waited until I couldn't stand it any longer, and I screamed out, "Shut up you assholes!" The quiet returned, and the night belonged to dreams and nightmares.

***

The clouds are gathering for a big storm. The wind is whipping the trees and snow into big piles as the cold air blows through the open door. I glance out of my bunk and watch as the figure entering the hut walks slowly toward my bunk. I froze in fear as it stopped by my bunk. The huge parka covered the person's identity. I was so alarmed that I couldn't move when the hand slowly reached out and began to slide under my blanket. The touch was warm and soft. My eyes became accustomed to the dark and peered into the parka hood, and the familiar face of Mindy Chambers stared back.

I couldn't believe it. She had found me. I couldn't move because I was sleeping at attention. Mindy caressed my body with her hands. I almost died when her soft fingers moved across my erection. She lingered there, and the pressure to move out of "at attention" was almost impossible to control. She slowly started taking off her clothes. I didn't dare move my head because that was against military regulations if you are sleeping at attention. I kept looking as far left with my eyes as possible. She was ready to remove her blouse. I saw the white bra and the tan color of her breasts. They were as big as I remembered. The bra had a front snap, and I couldn't breathe as she undid the top clasp and struggled with the bottom one. She was having trouble, and I wanted to help. I couldn't move though because I was sleeping at attention.

The sight of a huge black hand reaching over and undoing the final clasp was heartbreaking. I watched as the Black sergeant who had beaten me in the receiving barracks slowly removed the bra. Mindy's young breasts bounced as the harness was removed. The nipples stood out against the cold, and the indented bra lines disappeared as the sergeant rubbed them. Mindy was moaning and calling my name, but the sergeant explained that I couldn't move while I was at attention. He told her he would take my place. Mindy called out my name for help, but all I could do was lay there. The last image I had was the ugly sergeant smiling at me as he covered up my face with the sheet.

I lost myself to the pleasures of sleep!

LETTER #5

```
Well, we are still alive! I don't think for very
much longer. The D.I. says it is his personal
pledge to kill every one of us. He says he plans
on killing us over and over, whatever that means.

We live in a Quonset hut. We have the inside full
of double bunks. We have a wooden box at the ends
of the beds. A guy from New York is above me.
He's black. I never met a black person before. I
like him, but he sure talks funny.

We have to go about a block away to the toilet.
(Here it is called a head. I don't know why, some
```

kind of sailor talk) The same building has showers. They don't have any bathtubs though, so we showered this morning. We can't use hot water. You have to stand in the cold showers and sing the Marine Corps Hymn. The D.I. makes us all line up, so he can check behind our ears and stuff.

He said I'm too tall, so he puts his finger in the soft spot in my throat and makes me kneel.

We also wash all our clothes in this building. We all have to scrub our clothes on washboards. We then hang them out to dry. We have to guard our clothes until they dry. The D.I. said the sailors sneak over and steal our underwear. He said the sailors like to wear a Marine's shorts. I don't understand! The D.I. makes us do exercises while we guard our clothes. This place is weird. The sailors have boot camp across the water from us. I can't believe they would come over here to steal clothes. They are too lazy. We have already been up for hours before you hear them waking up over there.

LOVE, DALE

CHAPTER SIX

LETTER #6

Well, we got to go to the movie last night. The officers were all there to welcome the new recruits. Our D.I. was really mad. The officers all left when the movie started, and as soon as they were out of sight, our D.I. made us stand at attention, do an about face, and stand backward until the movie ended. He said that some of the scenes would poison our morals if we watched. He's crazy. The movie was "Old Yeller!"

We got back to the hut area, and he made us do squat jumps until bedtime. He said that the long time we spent standing made us stiff and sore! He was only looking out for our health and welfare. Vacky and I stayed up after lights out.

```
We were planning a way to escape. The fire watch
was patrolling the streets, so there was no way
out there. We might wait until we are marching
close to the fence and then make a run for it.
We thought of a tunnel, but our Quonset is right
in the middle of the base. We also stayed up
because we found a cigarette butt. We couldn't
light it, but we chewed the tobacco. It was
almost as good as a smoke. I don't think the D.I.
plans on letting us smoke. The guys are trying
to talk me into asking. I wasn't raised by no
dummy, even though the D.I. says my father has
wool. I plan on keeping a low profile. I get in
enough trouble without asking for it!!

LOVE, DALE
```

***

Dreaming

It's Saturday night, and we are done for the week. Uncle George and Ella are ready to go to Milner North Dakota. This is the highlight of the whole week because we've been bailing hay ten to twelve hours a day and custom bailing for two of the neighbors, in 100-degree heat. My brother Don and I have been here at Uncle George's since school let out two weeks ago. Uncle George has us stacking the bales behind the Baler on a stone boat he pulls behind him. It is an ingenious device. It is made of 2x6 slats, closed at the front and open at the rear. This allows us to stack the square bales with the bottom twines turned sideways, so the strings won't touch the ground and rot. The rest are formed into a pyramid that reaches 6 feet high. The top bale is placed to keep out the rain. A four-foot crowbar is then rammed into the dirt as the baler keeps moving, and it takes both of us to hold it as the pile slides off the rear.

We are back at the farmhouse. Ella has the steel tub filled with hot water. Don is first, and the dirt from the hayfield is thick coming off. When it's my turn, the water is now cool and dirty but still feels good on my 11-year-old body. Dressed up in fresh bibs and white shirts, we go off to town.

George has a special place to park by the Allis Chalmers dealership. He pulls the green Studebaker into the spot that looks over Main Street. Don and I are each given a quarter. George says to come back to the car when we are tired and sleep until he and Ella are ready to go home.

We first stop at Slim's Gas Station. We both buy a cola from the pop machine, where you slide the bottle over to the discharge hinge, deposit a nickel, and lift it out. We also buy a bag of Planters Peanuts that we dump into the coke for another nickel.

We meet up with several other kids and all go to the movie. It costs a dime to get in and five cents for popcorn. After the movie, we all walk around Main Street watching the traffic. We already know that Uncle George and the other farmers are in the bar, and Aunt Ella is in the restaurant with the other wives. We put all the windows down on the Studebaker. The night breeze is wonderful as we fall asleep from exhaustion to the ongoing sound of traffic as the night goes on.

I am awakened by singing, loud, out-of-tune singing. Uncle George is three sheets to the wind. He climbs behind the wheel, and off we go, over the gravel roads, sliding back and forth! Ella is scared and hanging on, but we think it is hilarious!

What happened? Did we wreck? Did Uncle George drive into the ditch? The glaring light that has come on amongst the shouts isn't George! I realized it was reveille at Marine Corps Recruit Depot, and I was dreaming!

LETTER #7

```
Well, here it is, another day gone by. Boy, we
used to think we had it rough on the farm. This
is what we do on a normal day. This morning, we
got up at 4 AM. We fell out on the street in
front of the Quonsets. We had roll call and then
to the huge asphalt, lot called the grinder. We
did exercises for a long time and then went to
the showers. We have to stand at attention until
our turn in the shower. We have five minutes to
have a cold shower, shave, and use the toilet.
We then fall back into formation.

We run to chow. I usually throw up on the run
back. I suppose it is because we eat so fast. We
return to the huts to get inspected, and we never
pass. The D.I. will put us on the street and make
us do squat jumps until everyone is on the deck.
```

One D.I. has a quirk that he performs every morning. He walks down the rows for inspection. He hits some of us in the stomach. He yells and cusses if we fall. He calls us maggots. He says it will make men of us. I have black and blue stomach muscles!

Then we go to class, come back, and march. When I say march, I mean really march. You get so tired; you think your feet will fall off. We go back to our area and clean, and then we head to the obstacle course. The obstacle course is something. You rope walk, swing from a rope over water (I fall in sometimes), and we go to the sand pits with our buckets. We do exercises with buckets full of sand.

We return to our Quonsets, go to supper, and if the D.I. feels like it, we can smoke half of a non-filter cigarette. We have to field strip the butt and place the paper in our pockets. We work on our gear. It is all becoming second nature now!

LOVE, DALE

The trip to the dentist was a whole new play on the words "gotta be tough." We were all lined up waiting for our turn to get in the dentist's chair. As my turn arrived, I was taken to a chair, and a dentist was assigned to evaluate me. They found that I had a few teeth that needed fillings and were discussing my wisdom teeth. It was determined that the wisdom teeth were okay, but they had to pull one on the top right side. I was waiting for them to use the needle to apply the Novocain. Instead, a couple of interns came in. One placed his hands on my arms and the other on my head, and the dentist came in and proceeded to pull the tooth without any shots. Luckily, I wasn't able to grab the interns or the dentist by parts of their anatomy to show my frustration at having that tooth pulled with no Novocain. They scheduled us for other work needing to be done, and we were put back out to stand in formation.

It seems that nighttime is the only time that we have to ourselves. So consequently, I guess I'm having more dreams than I ever had in my life. In one sense, this is good because it takes me away from reality and puts me in a fantasy land.

The daily life in the Marine Corps is meant to start with breaking you down to a zero position and then rebuilding you into the type of warrior that they want. The harassment and the things that look like abuse are actually designed to build you back up as a person. You are taught (or trained) to improvise, adapt, and overcome all obstacles that are placed in your way, anything that could otherwise keep you from doing the job you've been assigned to.

LETTER # 8

```
Well, I got my first taste of military medicine.
Our whole platoon went to the dentist. We have
58 from North Dakota, and the rest are from all
over. I even met someone from New York.
```

Well, back to my story. I got three teeth filled, and my head feels like it will explode. The Doctor said I was tough and didn't need Novocain, but it hurt like hell!

I forgot to tell you that we all got haircuts when we first arrived here. The barbers just set the razor on top, and all was gone!! I remember standing outside waiting and didn't even realize my best friend Phil was standing next to me! We all look completely different bald!!

We start taking self-defense classes and tests on the 15th. Right now, all we do is run, exercise, march, and more marching.

So far, I have only caught hell about a million times. You can't do anything right. I get out of step marching, not standing straight, standing too straight, moving my eyes at attention!! I hate to have to go to the D.I. Quonset. You have to run to the door, knock, wait for him to yell, "Enter Scum," then take three steps. You have to do a left face, look straight ahead, and say "Sir. Private Dallman reporting as ordered, sir." You are so nervous that you never get it right. Then he jumps up, stands right in your face, screams and hollers, spits in your face, and tells you to go out and do it again. I tried it 4 times, and he finally made me do squat jumps while the rest of the platoon tried. I was supposed to watch and listen but do squat jumps at attention! That didn't make sense! Pretty soon, there were about 20 of us doing squat jumps all over the floor. He would get mad and yell if

```
we bumped into anything.
That's all for now. Things aren't going too bad.
LOVE, DALE
```

Dreaming

The bed that Don and I are sleeping in on the farm at Cayuga, North Dakota is very soft. The Goose Down blanket that covers us keeps the frost from getting into our bones on a chilly night. We are up in the attic, and as you look up at the ceiling, you see the nails that jut through the pine boards. They shine with frost. The small window is covered with ice from the disparity of cold and warm air meeting right above our heads! We were so naïve as small children! We didn't know this was not the norm.

I can hear a noise downstairs, and I know what is happening. Grandma Dallman is up, way before the sun, and I can tell by the sounds that she is kneading bread that she made from scratch. I sneak down the sharp steps, open the attic door, and look into the kitchen. The smell is overwhelming. The fresh loaves of bread are sitting on the table wrapped with a cloth, and the wood-burning stove has the oven fired up and baking more. When Grandma Dallman sees me, she tells me to get back to bed. But before I scamper off, she quickly hands me a hot roll that has just come out of the oven. The taste is scrumptious, and she laughs as she swats me on the butt, sending me back up into the attic.

I crawl back under that goose-down blanket, savoring the taste of that hot roll I brought with me, and slowly drop back off to sleep.

I can't believe it. I hear that sound downstairs again. Grandma Dallman is opening the attic door and calling softly, "Dale come on down. I've got fresh caramel rolls ready for you."

My mind wakes up and then my body. It isn't Grandma Dallman. It is Corporal Val, and he is banging on the racks yelling, "Get out of the bed, you maggots."

LETTER # 9

Well, I'm sitting on the can, writing this. We just finished taking tests on General Orders, Rank Structure, etc. It feels pretty good to sit down, even if it is on the toilet with about 30 toilets in a row. The guy sitting next to me is from West Virginia, but he enlisted in Montana. He is a real hillbilly. He reported to boot camp wearing his dad's old uniform! Everyone makes fun of him, but I like him. He isn't phony like the guy from New York. We just came in from the grinder, where, believe it or not, we did push-ups, squat jumps, sit-ups, side straddle hops, ran around and around, and then got serious again with squat jumps. I'm glad to be sitting anywhere! I plan to flush and hit the sack!

Well, I'm back at the hut, waiting to go out on

the street for roll call. We got another haircut yesterday, though it was not much to cut. We have this suntan strip at the hat level, and it looks funny! My right knee is swollen, and I'm being sent to sick call tomorrow. We went to see the Marine Corps Birthday Parade on Thursday. The D.I.s made us do squat jumps as we waited for the cake-cutting ceremony. He said that was to keep us limber for the evening run he had planned.

Well, please write, and let me know how deer hunting went. Did my baby brother get a deer? Also, call Peggy to see why she hasn't written. Is she mad at me?

LOVE, DALE

Dreaming

The perfect day is dawning! I am at the park in Burlington, North Dakota. I just washed my 49 Ford. It was in Wahpeton, North Dakota with my brother Don. He drove it home from college where it was in the shop class for the last four months! They completely overhauled the Flat Head V-8. They bored it out, added duel, four-barrel carburetors, a 5-speed LaSalle floor-mounted transmission, and it was a monster!! They completely stripped down the body, chopped and lowered it, leaded in the door handles and put electric buttons in the grille to open the doors. They added a Continental Kit, fender skirts, and a Sun visor! They painted it a metallic blue and put in exhaust cut-offs so that with the pull of a cabled handle, it would go directly to straight pipes!!

I had Peggy, my girlfriend, and her sister with me. I met Peggy at the State Fair in Minot. We became a thing immediately. She was a bank teller at Union National bank. Her sister was a nurse at Trinity Hospital. Both were much older than me. I was still in Minot High School. She lived in Burlington, and her father hated me!

I was a gangly, kid, with long greasy hair, the Elvis Pressley style, in a Duck Tail. The black leather jacket, the engineer boots, and the huge Western buckle made him see me as a not-so-good choice for his daughter.

***

I am roughly brought out of the dream, as a trash can thunders through the Quonset with the D.I. yelling reveille!! It is 04:00, and the reality of my location is suddenly clear again! The fight to get outside begins as the D.I. wants us trampling each other on the way out!! We re-enter and re-exit about a dozen times!

There is roll call; then the mad exit to the 3 s! The showers are nothing but cold water, and we stand at attention and sing the Marine Corps Hymn.

I think *this isn't so bad.* I am one of the tallest, so I am usually first in the formation. The little D.I., Corporal Val inspects the formation.

Without warning, he punches me in the belly, knocking the wind out of me. I gasp as I hit the ground and throw up what little I have in my stomach. He continues down the line, doing the same thing to everyone. He leaves a mass of bodies on the ground throwing up or gasping for air in his wake. The only good thing about it is that the platoon is big, and he leaves me there for some time as he continues to torture the rest.

I closed my eyes and let my mind wander back to Peggy. I feel her by my side, helping me breathe. I suddenly feel better. The noise of the others is drowned out by Peggy stroking my long, greasy hair—BUT—Something is wrong!! I don't have any hair!! They shaved me bald! Peggy looks at me, but she doesn't recognize me. She runs to Phil and then to Vern, asking, "What happened to Dale?" I think I am going nuts, just as the D.I. comes back to me!

We fell back into formation and went on to chow! We got a new experience that day. We got to go to a huge field of sand! We were all spread out and holding our galvanized buckets in our hands.

The exercise was quite simple. You take your bucket, fill it one-third full with sand, and then hold it in front of you with both hands as the DI counts. We held this bucket out until the D.I. said "Stop." The weight of the bucket with sand started to hurt almost immediately. The D.I. looked on as we struggled to keep it out in front of us.

Soon, some recruits let the bucket drop. Unfortunately, this was what the D.I.s were waiting for. They suddenly appeared from behind and started yelling and pushing the recruits that let their bucket drop. This torture went on for what seemed like forever. They then instructed us to place the bucket on the ground and run in place to limber up!

We left the sandpit and continued to the obstacle course. It was a series of log-constructed structures used to train your upper and lower body. It included the most notoriously high rope climbs that reached up several stories. We had the option of either repelling down this wall or climbing a series of ropes that ran from the top of the structure to poles out about 100 feet that were suspended over a pond of putrid water. Most recruits who first attempted this fell into the water and had to walk around with that smell on them for the rest of the day.

***

I got assigned to head duty, which is a nice way of saying to clean the toilets. The building was separate from the Quonset huts we lived in and more than just our platoon used those heads.

Cleaning the heads was quite an experience since you scrubbed the floors and around the toilets with toothbrushes to make sure it was sparkling clean. I got this assignment because the D.I. decided that a few of us needed an "attitude adjustment," and being from the farm in North Dakota, we must be used to the smell of shit!

A cold shower later, and we were in formation to go to the obstacle course again. This time, I was on the swinging rope and dropped in the putrid water, so I smelled like a river rat the rest of the day.

A unique experience that we witnessed was the arrival of

"Pogeybait" (this is any kind of unauthorized sweets) via the mail to recruits. It is an experience that you do not want to happen to you.

Vacky was one of the recipients of such sweets. He received a huge box of chocolate from his girlfriend back in Minot. I think it might have been the worst day of his life.

LETTER # 10

> PLEASE, PLEASE, PLEASE don't ever send me anything but mail. DO NOT SEND FRUIT, CANDY, NUTS, POPCORN, or anything other than paper! Only paper letters. Don't even put perfume on the letters. PLEASE. Tell anyone who knows me not to send anything. Tell Peggy not to bake me anything, PLEASE!
>
> Vacky and a guy from Chicago got packages today at mail call. Vacky got a big box of fudge from his girl, and the guy from Chicago got popcorn balls. The D.I. made each go into the Quonset and get his galvanized bucket. He had them run to the head, fill the buckets with hot water, and run back. He made them go three times because they kept spilling the water. He then took a box of salt and dumped half in each bucket.
>
> The street in front of our Quonsets is pretty long. The D.I. had a recruit stand at each end. One held the water bucket, leaving the other

holding the sweets. They had to run to the one with the sweets, eat as much as they could, run to the guy with the water, and drink as fast as they could. They had to keep this up until all was gone. They started throwing up right away! It just got worse.

We start bayonet training pretty soon. The D.I. said each of us would try to stick him with our bayonet. He said if we missed, he was going to stick us.

I think he is nuts!

One minute, I want to get out of here; the next minute, I wouldn't quit if I had to. I can't believe I am starting to like being in the Marines. We really are the best.

We went to a parade, and the Navy and Army were there. They march like a bunch of pansies. They have no pride. We go to the rifle range pretty soon. It is at some place called Camp Mathews. They tell us we will be living in tents.

CHAPTER SEVEN

The rifles assigned to the Marines were picked up at the Armory. You were to inspect them to make sure they were clean and in working order. (they had never been cleaned, and since we had never seen an M-1 rifle before, we didn't know if it was in working order). We returned to the Quonset huts, and the D.I. showed us how the "field strip" the M-1 (meaning take it apart, clean it and resemble it). As the D.I. was educating us on the proper handling of the rife, he was adlibbing comments of wisdom. He said, "This is your rifle. There are many like it, but this one is yours. You will always care for it. It will protect you, and it will kill the bad guys. You must take better care of it than your wife, girlfriend, your North Dakota maggots, or your sheep at home."

He said, "It is better than a wife, as it can't talk back. It don't care where you go or when you return. It won't care if you have other rifles, and when it quits working, gets old and ugly, or you want a new model, you can turn it in for a new one."

He then went on to explain that we needed to get to know that rifle better than anything in our lives! We would tear it down and re-assemble it, over and over, until we could do it blindfolded. (we did get to that point). Later on, in their oblique sense of humor, they would blindfold you and hide a piece or give you double! The M-1 was a semiautomatic rifle that had a bolt that slid back and allowed you to place a clip, with several rounds in it, by using your thumb to hold it back under heavy spring pressure. If you didn't do the loading sequence in perfect unison, the bolt would smash your thumb against the receiver and give you a cut or broken thumb. (This was widely known throughout the Marine Corps as an M-1 thumb)

LETTER # 11

```
Well, I just got back from the head. I got a few
minutes before we fall out. You have to stand at
attention every time a D.I. comes in the head. I
was sitting on the john, and at least 5 D.I.'s
came through. I guess it teaches us muscle
control.

We had a Friday night smoker last night. This is
where you get in the ring and beat each other
senseless!

Most of the time, you fight another platoon. They
march you into a room and fight whoever is
```

standing in front of you. They line up the platoons from the tallest to the shortest, so most of the time you are fighting someone about your size. Well, last night the D.I.'s must have felt funny. They marched in Plt. 2003 and then had us, Plt. 2002, reverse our position. We marched in with our smallest first. I was the last one in. I did a left face, and there stood a guy who must have stood on his tiptoes to get in. We looked like "Mutt and Jeff" when we got in the ring. I kept putting my hand on his head, and he kept windmilling! In the second round, the D.I. came over and whispered into my ear that if the little guy survived another round, he was going to tie my hands behind my back. I believed him.

Allyn got a great big guy. This guy came out hopping and dancing around all over. He did this for about half a round, jabbing Allyn in the face, and making him turn red. Everyone was laughing at Allyn because he looked like a clumsy farm kid. The fancy kid bobbed and weaved and started dancing in front of Allyn. Allyn hit him once. They had to get the stretcher for the dancing kid.

LOVE, DALE

LETTER #12

Well, I only have a few moments but will write

until the D.I. hollers at us. Say, let me know what happened at deer hunting. Sounds like someone used the wrong tag. Who's in trouble? You people can't survive without me! Ha Ha

I goofed up yesterday and called my rifle a gun. I don't know how to put this delicately — The D.I. made me unbutton my pants, take my privates in my left hand, my M-1 in my right hand and run up and down the street yelling, "this is my rifle, this is my gun, this is for fighting, this is for fun. I had to do this in front of everybody. He would run alongside me, and when I got tired of holding my RIFLE over my head, he would yell and kick me. He made me keep this up until I finally dropped. He then asked me if I would remember what a RIFLE was. I assured him I would. He made me do squat jumps while the rest ate supper.

We have it figured out. We only have 66 more days until we are out of boot camp. Then only 60 days at the ITR (Infantry Training Regiment), before we come home (something like that) That's about four months!! I don't know if I can live that long!

We take tests to see what we will be. The D.I. had us all out on the street last night. He said he would light the smoking lamp (permission to smoke) for any that were planning to volunteer for the infantry. We all raised our hands. We hadn't had a smoke in weeks.

We all have to buy Pall Mall smokes. The D.I.

```
keeps them for us.
LOVE, DALE
```

The rifle range was both exciting and one of my favorite places in Boot Camp. I loved the M-1 rifle. It was hard for a lot of recruits who hadn't fired weapons, but my training on the farm was paying off in spades!!! I was good, no, not just good, I was at home with the art of firing.

It came so naturally. The M-1 is a fine rifle, but many who used it can attest to the fact that I got a "M-1 THUMB" (this was caused by not pulling your thumb out quickly enough when you changed clips). The range is set up so that you fire at targets in the Butts (this is a pit, behind the targets). Recruits are back there raising and lowering the target as well as marking the bullet holes. The score is relayed back to the firing line with colored discs that ease up. (the colored disc no recruit wants is called, "Maggie's Drawers"). It is a red sign that they wave back and forth to indicate you missed the target.

LETTER # 13

```
Well, how's everything in good old North Dakota?
It is raining here. We, the great U.S. Marines,
are out there marching and doing exercises. The
D.I. checks us every hour to see if we are
melting. I thought only people from Texas melted
in the rain!
I didn't go to church this morning, so I could
```

have time to catch up on letters. The D.I. doesn't care if we go, we just have to tell anyone who asks that we did. If the officers come through, we lie.

Tell Dad that the Marine Corps has given me a bed partner to sleep with. I curl up every night with my partner. It feels good to have someone to sleep with. My partner's name is 1664265, commonly known as the Garand M-1 rifle. The D.I. makes us sleep, shower, and go to the toilet with it! He says we must treat it better than we would treat a woman or our sheep back home. He says it will save our rear. He says it is better than a woman, and it doesn't cost anything to keep. It doesn't talk back, doesn't care where we go or what we do, and if we get tired of it, we can trade it in for a new one.

Things are changing and new every day here in boot camp. I hope I can make it, but… A Dallman can do anything, right? We graduate in January and then go to Camp Pendleton. Then, we get to go home on leave!! (I hope.)

They have given me more shots than you can imagine.

I don't like needles. They line us up, and a Corpsman gives you shots from both sides. I think they try to make it hurt. They also use an Air Gun to inject shots into us. The guy in front of me moved, and the blood was running down his arm.

LOVE, DALE

CHAPTER EIGHT

LETTER #17

Well, today is Christmas. It sure doesn't seem like it though. The D.I. told us that the Commandant of the Marine Corps has declared today a holiday for all Marines. Nobody has to do any work, exercise, or anything that isn't fun. The D.I. said he was going to take us bird watching after the noon meal.

We went to chow. They had some girls at the mess hall to sing Christmas songs for the officers. I couldn't keep my eyes off them. The D.I. walked up behind me and said I might have to visit him after we get back from birdwatching. I ran into the guy ahead of me because I was watching the girls. I suppose he will chew me out later.

We just got back from birdwatching. The D.I. has

us fall out in the street in boots, packs, and rifles. He said we were going to go out in the country to watch birds. It was strictly volunteer on our part, but he wanted 100% participation. He said the reason for the packs was to carry back any wounded birds we might encounter. The reason for the rifle was to protect us against the horrible attacking Sparrow!

He said the best way to watch birds was to move along briskly and see as many as possible. We ran from 12:30 until dark. We all threw up our Christmas dinner. The D.I. said he was just following orders from the Commandant. He took us on an educational excursion. We qualify on Friday and then hike back to MCRD.

LOVE, DALE

LETTER #18

Well, today is Tuesday. We have been firing live ammo. Most of our time up to now was "dry firing." This means assuming the position, but no ammo. Kind of like a steer! Well, now we are bulls!! You can't believe the thrill of feeling the kick of your M-1 bed partner!

We shoot 200 yards offhand with a sling, 300 yards kneeling and sitting, and 500 yards prone. The targets are a 10-inch bull and a 20-inch dog target. The best score you can get is 250. So far, I have been recording my sight settings in a little book called our dope book. I have shot a 188, 223, 221, and a 230. The only one that counts is qualification day!! I hope I don't panic and fall on my rear.

We leave for MCRD right after the qualification day. We have to hike or run back.

We all have to go to church. It is a custom to receive a prayer before you qualify. I guess it makes you shoot straighter. I don't know if GOD cares if we shoot straight or not, but it can't hurt to believe.

I got in trouble because of the girl singers I looked at, during Christmas. I'll write about that later. I sure seem to like the females. I wonder if that will be my downfall!!!

LOVE. DALE

P.S. Nothing from Peggy!! She must be mad at me since I didn't see her when we left for boot camp! I meant to, but we got carried away!! You can tell her I'm sorry if you see her!

Reminiscing on these two letters brings a moment of joy, inspiration, good memories, and bad (memories). Being at the Rifle Range at Christmas, was without a doubt, one of a few things that were completely out of wack.

Christmas in Minot, North Dakota was celebrated as a town. Main Street was aglow with lights and a real Christmas Tree in the center. Picture a Norman Rockwell painting, and you see

Minot!! The snow was always deep, the air cold, and the people happy. If you went into stores like Ellison's Dept. store, you'd find decorations, Santa Claus, and elves!

A sleigh with beautiful black horses took riders around the downtown areas. They used huge Buffalo Robes to cover the riders, and the generous supply of Egg Nog (some with Rum) kept the dreams and romance flowing.

Dad always went out and cut a tree in Montana when he was an engineer on the Great Northern Railway. His Conductor was usually Henry Ruff, so both of them would put the trees in the caboose and smuggle them back to Minot. It was an annual event, and both families would help each other decorate.

This Christmas was to be the first that did not fit anything I remembered! It started as a declared holiday, but typical of the Marine training, became a nightmare! We ran for miles. The Christmas dinner was left on the dusty trail as we "bird watched."

This brings me to the fact that while we waited in the chow line for that "special dinner". A bunch of girls were brought in to sing Christmas Carols for the Officers mess.

They paraded them right by me!! We were standing at attention, but I couldn't resist. I moved my damn eyeballs, and the D.I. saw it! I knew I would pay the price. I just liked women, way too much!

***

This all made me reminisce about the first Christmas with

my girlfriend Peggy. She always wanted a Cross pen. It was something I could never afford while sacking groceries at Piggly Wiggly. We had a customer who was a doctor's wife. She always shopped on Saturday. All the other kids tried to avoid her as she checked out because she had a full cart.

It was a big chore to help sack her groceries and then take them to the Studebaker Station Wagon she drove. I would always volunteer to help her. She was mean and demanding, so the rest were happy I took the challenge. She was so particular about how they were bagged and had to okay the exact placement in the back of the Studebaker. She never said thank you, but I would talk to her as I loaded them up. She always tipped a dollar! I began to take her as a challenge.

Five days before Christmas, I am loading her car and find she has traded the Studebaker for a big Mercury Station Wagon. I complimented her, and she told me her husband bought it for her for Christmas. I am rambling as I load the bags and mention my girlfriend has always wanted a Cross Pen. Nothing more was said as I put in the last bag.

She comes over, and I am expecting my dollar tip. Instead, she reaches into her purse and hands me her pen. I'm sure I looked as I felt, puzzled. She tells me to look at it. It is silver with gold groves running up and down, and in fine print, it says "CROSS."

***

It was a perfect Friday. I remember that we got up early to have a good breakfast. The D.I. walked us to chow and gave us a "Smoking Lamp is Lit." We got to the range, and the instructors and Officers took over. It was very organized, and soon we were firing for the official qualification.

It started with the 200-yard line offhand/standing position. You had a leather sling that you wrapped around your left elbow and sighted on a target that I think was a 10" Black Bullseye. You have recruits in the area called the butts. (they are in a trench, below the targets) Their job was to pull the target down into the trench, read, and paste the hole the bullet left. They had poles with discs on them. A white disc was a bullseye. A complete miss drew the red flag that was swung back and forth! (This was called Maggie's Drawers, and it was meant to shame.)

Next, we did the kneeling and sitting positions at 300 yards. The last was the prone position at 500 yards. We had a little book, called the Dope book, which we recorded the sight settings in during the week. It would give you the proper settings for each yardage. You had to figure out the wind and adjust for that.

My weak spot was the offhand standing position at 200 yards. I had learned how to breathe, so that I could fire as my rifle was coming down into the black bull area and keep the motion. I dropped most of my points at the offhand position. I did very well with the kneeling and sitting positions, and I think I only dropped one or two points there. The 500-yard line was my perfect score. I dropped all the rounds into the bull's-eye, and that qualified me as the high shooter of the Platoon and the series.

Since it would be illegal, I don't want to say it happened. If it did happen that I shot the 500-yard line for a couple of other recruits, at the request of one of my D.I.'s to help the team, that would make a great story!

LETTER #19

Well, here it is, Friday afternoon. We fired for qualification today. I did ok. The range was tough. The art of firing a large caliber weapon at a small bull was not an easy feat for the faint-hearted.

I used every ounce of energy my body could produce. I thought of my love of family, country, *Espirit De Corps*, and the fact that my D.I. would kill me as I pulled the trigger of destiny.

I (your humble son) fired a good score! (227) It was the high score. I will receive a trophy from the Marine Corps. Phil, Vacky, Virgil, Vern and Dick also qualified.

We have to fall out for rifle inspection in one hour.

The rifle has to be really clean, and I mean *really clean*! We take the rifle in the shower and use hot water when we can. They hold it up to the sun, and if they find any oil, lint, or dirt, you do squat jumps for the rest of the day.

Did you hear about Red China and Vietnam? It

sounds like the Marines are going in, in secret. This would be great. We want to fight. Hell, all I have to do is hold up my expert rifle badge, and that would scare the hell out of them!

We get our clothing issue when we get back to MCRD. I mean real uniforms, not the utilities we wear now. We get shoes, shirts that have creases, and pants that we don't have to use blousing strips on. We will finally look like real Marines.

The D.I. made me report to the duty tent because of the singing girl incident on Christmas. He had the platoon fall out in the street in front of the tent, and then he lit the smoking lamp. He called me in. To make a long story short, three D.I.s worked me over for about an hour. One of them threw me right through the side of the tent. I knocked over the stove and landed right outside in the formation. He yelled for me to come back in. They told me they were proud of my high rifle score as I did squat jumps! I got some sore ribs, but they can't hurt me anymore. I'm a Marine!

LOVE, DALE

It seems that I remember the Leatherneck magazine participation and the bond participation taking place at payday. We got paid in cash in those days. They would set up a long table, and an officer and another one watching over his shoulder would count out our pay in real money. Our pay was about sixty dollars a month.

***

I thrived on the Rifle Range! I remember it now, 64 years later, like it was yesterday. I was in charge of my destiny. The Marine Corps and the D.I.'s couldn't control me. This was the one spot where I was in charge. I was a natural. I fired a good score and was top for the Platoon. It was interesting to see that all the North Dakota recruits did very well. Most had been brought up with firearms. It was a common thing to hunt, fish, and plink tin cans! This wasn't the case with the city boys. Many had never even fired a rifle.

The run/walk back to San Diego was not as hard as I imagined. We got fairly close, and some Cattle car trucks loaded us up and took us through the city.

I was called into the duty hut, and I went through the agonizing ordeal of pounding on the hatch (door). I waited until the command, "Enter Maggot," was called out, and I opened the door. I took four steps forward, did a left face, and in a loud voice said, "Sir, Pvt. Dallman requests permission to speak to the Drill Instructor, Sir." I looked straight ahead, but my peripheral vision picked up several people in the hut. A photographer had been sent by the Congressman from North Dakota to take pictures of me because I won the "High Shooter" Trophy, as well as the "Outstanding Man" Trophy for the North Dakota Platoon. I was turned over to the photo team, and they took their shots.

This memory went hand-in-hand with another less-than-stellar one, which was usually the case in our platoon. The D.I.'s would find any way to keep us 'humble.'

The platoon had four squads and four squad leaders with a total of between 50 and 80 recruits. The leader's job was to be in charge of (take the heat for) the 15 to 20 recruits in each of their Platoons. If someone messed up in your Platoon, you have a couple of options: help them or do a late-night operation called Blanket Party. (this consisted of throwing a blanket over the malingerer held down by the rest of the squad as they used a sock with a bar of soap in it to "adjust their attitude") The other option was, if they did something wrong, the DI would have them stand at ease and give them a "the smoking lamp is lit". The Squad leader took the punishment as they watched! You can imagine that the Squad leader wasn't too happy and tried to instill in their feeble minds that he wanted them to never put him in that position again!

The first Squad leaders were chosen purely by height, then they were weeded out by attrition or failure to perform. I was appointed right away by height. As the weeks progressed, the squad leaders would change quite a bit. I only lost my position once, by being on the losing end of a Challenge! Fight, where I was knocked out.

Almost right after my photos were taken, I ended up losing my squad leader position. We went behind the huts and put the gloves on. The D.I. told me that since I was a Hollywood celebrity and the photo crews put makeup on me, he needed to make me not so pretty. He said I had to let the kid from Texas hit me twice before I could fight back. Well, there was no fight after that!

I don't remember too much about the Texas kid knocking me out for the Squad leader position. It would probably be the third or fourth time that I had defended my squad leader position. Getting knocked out should be in my brain, but it's not. I do remember that the drill instructor made me do squat jumps for a couple of hours to get me ready, so I could get my squad leader position back.

I never was relieved due to incompetence.

As I wrote a letter to the folks, I also penned one to my lost girlfriend Peggy. I tried to explain what happened before we left for San Diego. We had been partying, and I just didn't have time to see her. It was no big deal! I would see her when I got back, but I haven't gotten a letter from her for some time!

LETTER #20

```
Well, we survived the trip back to MCRD. Several
Platoons were together when we started, but we,
Platoon 2002, left the rest in the dust. We are
really in good shape. The D.I. said he was proud
of us. I couldn't believe it. He talked in a soft
voice and acted human. He was proud of us! We got
back to our Quonsets and he gave us an open
smoking lamp. I smoked two Pall Malls! He let us
walk to chow and gave us another smoke break! I
got dizzy because I hadn't smoked that much in
months!
```

It is now 08:30, Tuesday. We got issued our uniforms.

They are now being tailored. The Marine Corps is taking pictures of me for the newspaper in Minot because I was the high shooter. The D.I. made me do squat jumps, so I wouldn't get a big head.

I have about 20 days left here in boot camp, then on to combat training at ITR (Infantry Training Regiment) in Camp Pendleton. I hope I get a chance to get into this thing in Vietnam, even though it is a secret! It sounds like we might get a chance to use what they are teaching us. We train every day to be ready. Hope it's not over before I get out of here!

I ordered The Leatherneck Magazine today. It will be arriving there every month. We were told it was a volunteer thing, but our D.I. wanted 100% participation. I also took out a $25 savings bond per month. They will be sent to you. It was voluntary, but, again, the D.I. wanted 100% participation!

I lost my squad leader spot to a guy from Texas. He hit me a good one and knocked me colder than hell. I will get it back on Saturday!!

LOVE, DALE

P.S. Is Peggy dead??

CHAPTER NINE

The trip back to San Diego went quite well. I remember that we had a really good physical platoon, and we quickly outdistanced the rest of them.

When we got back to the Quonsets the D.I. let us smoke a couple of cigarettes, then later gave us another smoke break. I continued to get dizzy since we hadn't had that many cigarettes in months.

Life was moving fast, and so many things were taking place it made your head spin.

We started getting information, small pieces at a time, about the Marines secretly going into countries with names like Laos and Vietnam. It looked like it was going to expand to the point where they were probably figuring most of us would end up going there. It was the biggest topic of discussion) between recruits, and most understood that was why we were there. There was no panic, fear, or concern. Instead, most of us were eagerly awaiting the opportunity to get in the fight. We were already a group that had changed. Pride was now a driving factor. You were already instilled with the fact that the Team is what mattered. You were a cog in the team wheel. You are trained so that you don't quit on the team. You don't value your life over the team, but are willingly ready to give your life to save the team. It is quite interesting how the Marine Corps purposely breaks you down to nothing and then builds you back into the team concept.

An interesting thing happened at the chow hall. Our platoon was waiting to go in, and another platoon was standing nearby. Some quiet insults were tossed, and suddenly two recruits from the other platoon stepped over and hit one of ours! There was no hesitation from either platoon! All 70 plus recruits from each platoon were in the fight! The D.I.s were helpless to stop it. Soon, several M.P.s and other D.I.s were in the mess. It ended as quickly as it started. We were not punished, so I sometimes wonder if the D.I.s didn't plan it.

We also got issued our uniforms about this time. I think we got one set of winter dress greens, which was a coat, trousers, and two covers. One was the round cover with a bill, and the other was a wool cover that was generally known as a "piss cutter." We got some light tan uniforms that were basic. I think we called them Trops. We also got a pair of brown shoes. It seems I remember getting another pair of combat boots and some more utilities at the same time. The boots were old issue brown, so we had to dye and polish them to a spit-shine glitter in black. We used dye, polish, lighter fluid and spit! The boots were different than those in other services. Ours had eyelets that were designed so that you could quickly release and kick them away in the water.

LETTER # 21

Well, as of today, we have sixteen more days left in boot camp. I can't wait to get out of this hovel of misery. The D.I. has really been on the prod lately. He says we aren't tough enough; that we need to be like rocks. He says nobody will graduate, and he will be our mother forever. He says he will break every one of us. He says we are all sniveling maggots from North Dakota. He says all of us have sheep in our family. He is just mad because none of us fall when he hits us in the stomach anymore!

Oh, yes. I got my squad leader position back again.

The kid from Texas wasn't so tough, just a lucky punch. Nobody challenged me last Saturday. I hope they let me keep it. I will get PFC stripes because I was the high shooter and a squad leader.

We leave on a long hike tomorrow. We will be wearing full gear and weapons. It doesn't matter; we don't even breathe hard anymore! I can say one thing about our D.I.s; they have us in good shape. No other platoon can even compare. Hell, we could run all the way on this hike!

They have been working our butts off lately.

Yesterday we goofed up on the manual of arms. We had to do squat jumps until midnight. The officer of the day came by about 2200. He told the D.I. something, and we were put in our individual Quonsets. When the officer left though, they put us all in one Quonset. We then had to continue with the squat jumps. Most of us made it until he stopped. The few who didn't have to clean the heads for a week.

LOVE, DALE

P.S. Does anyone hear from Peggy?? I know I treated her wrong when I left, but she should understand that we partied hard before leaving. I just didn't tell her goodbye!!

CHAPTER TEN

The D.I., Cpl. Val, is really on the prod. He is the (little rooster) of the D.I.s! Seems he has a bit of a deranged personality. The others correct us and try to build us. The 'Little Corporal' is just plain mean. He is the one who liked testing his fist on our bellies. It hurts, but most won't let on.

I did get to win back my Squad Leader position.

The kid from Texas was tough, but as we squared off, I noticed something that defeated him before we started. There was fear in his eyes! He didn't have the confidence to win! It was over quickly.

I was starting to wonder about Peggy by this time. There were no letters! My fool of a young brain-didn't understand that she would take it so personally. We were all blowing steam, and the fact that I didn't see her before we left was not a big thing (in my mind)!

Now, years later, with a long string of female acquaintances, I understand that it is IMPOSSIBLE to understand a female!

My legs started going out. The front of them was painful. It was a deep pain that seemed to emanate from the bone.

It was hard to describe. It would hurt to start on a run and then it would become numb. I would wake up at night, almost screaming from the pain. (I found out that it was in direct correlation to the squat jumps. The very squat jumps that were outlawed soon after we graduated.)

The other guys were bothering me when I was writing a letter. They all had been constant guests at the Dallman house for years while we grew up. My mom was just as much theirs, so I let each of them scribble something.

I was already planning a way to cover up the hurt legs, as there was no way I was going to be set back in Boot Camp! I truly believed I would have died first!

Being set back was something that had happened to so many in our Platoon. They would go to a medical platoon to heal. The harassment never stopped.

When the person was signed back in as "ready for duty," they were dumped into another platoon. The newly assigned platoon could be any place in the training cycle! Since I was in the last week and could easily wind up in some platoon that still had weeks or months to go, I wasn't having it.

The squat jumps were the hardest on my swollen legs; they were turning black and blue. I had to be careful and make sure the D.I.'s didn't see them! The fact that we group showered was what made it the hardest to cover up. I would try to stay in the back of the crowd.

A lot of our time was now taken up in fitting uniforms, classes, and marching. The marches hurt but were better than the exercises and running the obstacle course!

Then a miracle was delivered! I was assigned fire watch. This meant I would walk the Quonsets all night to watch the safety of the compound. This made me miss many of the exercise and squat jump sessions.

LETTER # 22

```
Well, here it is Sunday. We got nine more days
in this damn crap joint. I'll never be so glad
again to get out of here. I have something wrong
with my legs, so I can hardly walk. I don't dare
go to sickbay. They will set me back. I know of
at least 12 recruits (or men) out of our platoon
that have been set back. I would die first. One
of the guy's cousins is a Corpsman here. Maybe I
can get some pills. I can't let on that anything
is wrong.

Here is Dick-he wants to say, "Hi."

"Hi. Boy, isn't Dale a crab? You know, ever since
we got here we have been counting the days until
we get out. But, to me, it don't seem that long.
```

But, boy, trying to convince Dale of that is impossible. It's about time for me to go. "Hi, this is Vern. You probably won't remember me. Dale wanted me to say a few words. I don't know what to say. You probably won't even know Dale when he gets home. He is nothing but an ox! I bet he gained 50 pounds here. He's the meanest SOB in the platoon. I can't think of anything else to say. Goodbye."

This is me again. Don't ask how I got some pain pills. I cannot start over again. Don't say anything. I love you and respect you, but this is my business.

LOVE, DALE

CHAPTER ELEVEN

It was wonderful to graduate. I got my legs to hold, and the graduation ceremony went very well. I did get my Private First-Class stripes as I was the squad leader and the high shooter for the platoon. I think three others got PFC stripes, none that I recall were from North Dakota.

When we got to Camp Pendleton, we were all spread amongst many different companies. I happened to end up in N Company, but I was immediately sent to Sick Bay. They took a look at my legs and pulled me out and sent me to a Causality Company. I remember the doctor looking at my legs and doing some X-rays. He asked me how long it had been that way. I told him it had just started. He just shook his head and looked at me.

It looked like I was going to be a month or two on crutches with some sort of a cast on my legs. It wasn't concrete but was more of a fiberglass cast.

This put me in a unique position in ITR, as I was not assigned to a training unit. Therefore, I had somewhat more freedom to go places on the base. I remember going to what we called a "slop shoot", which was a bar café kind of place. I also remember going in there, taking some pictures in one of those photo booths, and sending them on to the folks. This way, they could see what I looked like. I had gone from a 150-pound skinny kid to a 190-pound Marine with no hair!

I broke all the rules of being locked up for so many weeks in Boot Camp. I smoked, drank, and ate all the Pogey Bait I could find.

My daily routine was very slack. I would fall out for rollcall, make my medical appointments, and the rest of the day was mine! I got pretty good on the crutches and could find my way around to the Commissary and the Slop Shoot. I even got to go to a movie or two.

I am told about this time, that I am probably going to get to have liberty. This means, as soon as I can walk well enough, I probably get to go to town.

Interestingly, I never took the time here to try to contact Peggy by phone. I don't know why. I guess my new life was consuming my time. I suppose now that I look back on it, I was afraid to face the consequences of no contact with her when I left.

In reality, Minot, North Dakota was so far away. It seemed like a dream more than anything. I thought of Peggy, but honestly, I didn't have any strong feelings. I guess the crush on a girl at seventeen wasn't really what you would call love. I suppose all boys remember their first, but they don't realize until they're out in the world, that it certainly won't be the last.

Life was changing so fast, that the new horizons far outweighed the past. You take a farm kid from North Dakota, dump them in the middle of California, and it's like nothing you can imagine.

LETTER #23

Well, will wonders never cease. We graduated! I got PFC stripes, and I'm FREE!!! The cattle cars picked us up, loaded our gear, and headed for ITR at Camp Pendleton, California. We were received there by trainers who actually treated us normally. I understand we might even get liberty soon. It has been months since I was anyplace civilians were!

We are all scattered in different companies. I am in N company. There are 12 of us here that were in Platoon 2002. None of my buddies are with me. I met a guy from Britton, South Dakota, and he knows some of our relatives from Rutland and Cayuga.

I just got back from the "slop chute" (bar and cafe) where I had some pictures taken in one of those photo booths. I'm sending them to you, so you know what I look like if I ever get home again.

I have run out of pain pills, so it is sick call tomorrow. My legs are all swollen, and the fronts are black. It hurts when I walk.

I am out of boot camp. That is all that matters. I will be a good boy here. I promise. I want to come home, but if I'm delayed now, it is ok. I can live with it. I'm going to a flick tonight. I can even smoke when I want here. I plan on drinking beer tonight. I also plan on eating a hamburger, french fries, onion rings, candy bars, ice cream, and anything else not good for me!

LOVE, DALE

P.S. If you see Peggy, tell her I am sorry for not seeing her when we left. I will make it up when I get home if she isn't married or something.

It was only one of two times in my life that I had ever had broken bones, cast, or any type of invalid recovery time. The other was back in Minot, North Dakota. I was helping my brother put in pipelines in the Badlands around Squaw Gap, North Dakota. We were working for the water commission, installing stock tanks in areas that had no water available. We would put in miles and miles of underground water pipe and then bring it back up into stock tanks that were covered with dirt, which made water accessible to the cattle.

It was at one location that we had dug a large hole to put a pump station in, and I was down putting the last parts together as the bank caved in and buried me and one other worker. I was buried to my shoulders, and Larry, the other worker, was buried. I could only see his face. I worked one arm out and started pushing as much dirt as I could away. I finally got both arms out. Since Larry was just about a foot or so away from me, I started scraping dirt away from his face, so he could breathe.

I could tell that his breathing was labored because of the tightly packed dirt around him. I finally snaked myself out of the hole and realized that my left leg was completely broken.

By this time, my brother had come into the hole with a shovel and started digging around Larry. I was digging with my hands, and we finally got him loose and onto the side. Suddenly the remaining balance of the dirt caved in. It again buried the three of us, but not so deep that we couldn't fight to get out of the hole.

We were miles away from any type of civilization, so they loaded me in the back of a pickup truck and drove to Sidney, Montana, which was probably 40 miles or so away from our location. I still remember bouncing around in the back of the truck thinking it was a dream. We arrived in Sidney at the hospital, and they brought out a gurney, loaded me up, and hauled me in. They did some X-rays and put me in a temporary cast. They didn't have the equipment to work on the leg, so they called my wife in Minot. They had her drive 100+ miles to Sidney to pick me up and haul me back to Minot, to Trinity Hospital. I ended up with a really good doctor, and he had to go in and put some pins in that left knee. He put me in a cast, and I was laid up for the better part of the year. I remember my daughter, Jessica, was just a small girl at that time, and as I lay on the couch, she would cuddle up and try to protect me from the big, bad world.

This accident changed my lifestyle, and after many different endeavors and endeavors, I ended up in Billings, Montana years later. I was working as a salesman for a drug company, and one afternoon, Jessica and a friend of hers wanted me to take them out while running the dog. We went north of Billings to a huge open area that had large ravines running through it. This was the start of another adventure that I have written a book about called *The White Buffalo*.

We found a Buffalo Skull that measured out to be one of the largest ever found. The girls took it back to town, and the sequences of events that followed were earth-shaking!

The story of the White Buffalo is thrilling and provocative!

LETTER #24

```
Well, I'm writing this to all of you because I
don't know where I'm going to be. Peggy isn't
writing, so maybe you can forward it to her. I
am, for the moment, in Casualty Company at 2nd
ITR. I have stress fractures on both legs and
trouble with the arches. They are on the shin
bones. The damn things hurt like hell. I have
been put on no duty until March 2nd. The doctors
say I might be left on No Duty until April. It
takes four to sixteen weeks to get better, and
```

they never fully heal. It is directly caused by squat jumps.

I can't believe this. The Dr says he has never seen such a severe case before. He wanted to know how long they had been bothering me. I told him it just started. He just looked at me and shook his head.

I am on crutches. I have never been so p----d in my life. I was looking forward to going home on the 10th. It looks like I won't even be starting training by then. I asked the Doctor for pain pills, but he said, "No." It just covered up the pain, and I might really do further damage using them.

I've been doing pretty good until now. I kept trying to outperform and outdo everyone. There isn't anything I can do but lay around and think about it now. I don't like being helpless and depending on someone else. I'm not even supposed to go to the can by myself. I ain't taking nobody into the head with me!

LOVE, DALE

CHAPTER TWELVE

TJ!! It was a town on the border. We caught a bus to the city, then we arrived at the crossing. I seem to remember walking across a bridge. My mind is cloudy here! I think we got a taxi, but the bottom line is we arrived in the bar area. There were about six of us. We all wandered in and out of the bars.

Every bar is full of girls! The solicitations are nonstop! We are still kind of timid! Everyone wants you to buy a drink for them. We were warned by the people at the base, not to order any drinks that needed mixing. We were told to only order our beer in unopened bottles. The money exchange was strong for the American dollar! The girls would try to get you to pay a dollar for their drinks (most were tea). We got suckered a few times but kept moving. We hit a bar called "THE BLUE FOX"!! WOW!! They had a chain up, directing traffic, so that you couldn't enter without buying a beer. The interior was very dark, and as your eyes adjusted, you saw a large (HUGE) stage in the middle. The tables were scattered around, and a lot of Americans were sitting with the girls. (I'm trying to describe this in an acceptable way). The girls were involved in some practices that took them under the tables. We found a table near the stage! The first act that came out was a young lady who stripped, and then she practiced a stunt with a stack of quarters! It seems she could even make change! (I'm trying to keep it nice). Seeing her able to drop the exact change was dramatic!

The next act was the one that the "BLUE FOX" was world famous for—it involved a cart, a donkey, and a young lady.

This was one that I couldn't believe. It was about this time that our table was joined by several girls. They wanted drinks! We bought them. Then the haggling began for the world's oldest profession. Four of us decided to partake. They wanted us to go with them in a taxi. We figured, ok! We went out the back door of the club, into a taxi, haggled the price of the ride to ten dollars, and away the driver went! We were shuttled in and out of back alleys, up and down some streets, and finally arrived at this dirty old apartment building. We went in, did our 3-minute duty, and were back out. The taxi driver was there-he wanted twenty dollars to take us back! We had no idea where we were! We threatened him, but he stood his ground. We paid up and away we went! He dropped us at the front of the club. We gave him the bird as we got out.

We all decided to take a walk to another club, so we went to the end of the block. We turned left, walked maybe 100 feet or so, and suddenly realized that it was the old dirty apartment building we had just left with the girls. It turns out that our ride was a con game, and the place the girls took it was less than half a block from the club.

We had to laugh! We got taken in by the old taxi scam!

We spent the rest of the night in clubs drinking beer and finally ended up back at the border. We discovered that we could take a big, old, black taxicab that would take us all the way back to the base rather than go to the bus depot.

I had my first run-in with a homosexual on my last trip to TJ. It happened at a bus stop. That wasn't something we were familiar with coming from Minot, North Dakota. I'm sure they were there, and I remember one that owned a Drive-in restaurant. This gentleman was overly friendly, and the rumors were, he enjoyed boys. I never got the opportunity to meet him, but that would've been my closest encounter.

The only other encounter would've been with a friend of my brother's who was walking home from school and disappeared. His body was found many weeks later, north of Minot, under a bridge in the Souris River. It turned out that a man in a white foreign car had offered him a ride home, took him out there and killed him. He dumped his body with blocks tied to it under the bridge. It was a huge story in the Minot Daily News.

The trial was statewide news. If I recall, one of the pieces of evidence that they found in the man's car was one of the boy's shoes, which was a Blue Suede shoe, made famous by Elvis Presley!

The media was at the base looking for me. It seems I had won both trophies for the NODAK PLATOON. The Minot Daily

News had them come to take pictures of me for the hometown paper! It was funny because they had me in full battle dress, crawling over a wall. The M-1 Rifle is in full sight! They wanted me to look mean. The funny part was they had to help me up on the wall, as I was on crutches at the time!

LETTER #25

Well, I will be at the following address until I get back in training.

PFC Dale E. Dallman, Serial # -------

Casualty Company, 2nd Btn, 2nd ITR, MCB, Camp Pendleton, Calif.

I had a 72-hour pass last weekend. I spent it all in Tijuana, Mexico, a Mexican town just across the border from San Diego. WHAT A BLAST!! I did, saw, and witnessed things that I didn't know could be done! I spent my whole liberty there.

I can't go into descriptions here, but all the things the preacher back home said were bad, were invented in this Mexican town! He wouldn't believe what goes on. I LOVED IT!

I'm going back next weekend, and the next, and the next! One place that will always stick in my memory, is a bar called the "Blue Fox!"

I sure hope to get out of here soon. I want to finish up here, so I can come home.

I found out the reason the publicity men were here taking pictures of me. I guess I won both trophies for the North Dakota Platoon. I guess I'm a celebrity. They sent me a letter saying I won both the high shooter and outstanding man trophies. I guess I will get them if I ever get home.

I don't feel like a celebrity right now. I just feel useless. But--I guess if you got to be unless, Tijuana Mexico is a good place to forget your worries.

LOVE, DALE

PS You probably don't want to show this letter to Peggy.

My folks sent me the nice write-up! They played up the part of the fact that many of us who just graduated Boot Camp would be heading for a little-known country named Vietnam! The secret had been out for some time that Marines and Special Forces had been in the country for a while.

LETTER # 26

Well, we finally started training again. It feels good to be back in training. You can't imagine how long and high the hills are! You start running up-up-up!! They seem to never have a top!

The weather is worse in Oceanside than in San Diego. Even though it is just 25 miles up the coast, it gets extremely hot, and then cold. It's easy to freeze your butt off at night too.

Speaking of butts--I had my last liberty before returning to training. I sunburned my rear end. I can hardly sit down. A few of us and some Senoritas went to the beach for two days.

They are threatening to write me up for the destruction of government property!

You know, if you get to the ocean, this country is pretty good. I don't like the amount of people out here. You have to see the roads to believe them. The Ant Hill cities have six and eight lanes going in each direction. People drive like they are nuts!

I was on my way to Tijuana. Four of us caught a bus on the main side. It took us to a bus stop in the city. We had to change buses there. I walked into the restroom to use the urinal. I was standing there, holding my business, when suddenly, an extra hand showed up to help me! I wet all over myself, turned around, and found this guy smiling at me. I lost it. I hit him so hard, that he just dropped and started quivering. I ran out into the terminal and right into the shore patrol. I told them quickly what happened. They said to get out of here!! They protect those kinds of people in California. I said THANKS and got on my bus to the border. I hope I didn't kill him, but nobody grabs me there.

Well, maybe females can.

LOVE, DALE

P.S. Peggy sent a short letter!!!

CHAPTER THIRTEEN

I am ready!! The training started, and I am ready! All the fun and liberty is over! I am assigned to an infantry training unit here. This is the last hurdle before I get to go home on leave.

It is rather dim in my mind right now. I seem to have forgotten a lot of the time frame, but I will relay the things that shaped me for the future. The real side of our training began here!

The months in Bootcamp formed us into disciplined Marines, but this was where you learned the scope of death and destruction a Marine Combat unit could inflict on the enemy.

We covered the vast array of weapons. We fired the B.A.R. (Browning Automatic Rifle). This was a heavy rifle, approx. 20 plus or minus pounds that held a clip of 20 or 40 rounds. It was a gas-operated light automatic weapon that was rated at 500 rounds per minute. It was designed to be a fire-on-the-move weapon.

We also got to fire the 30-cal Machine Gun, which was an air-cooled weapon on a tripod. It was a true machine gun, capable of many rounds per minute. We also had the privilege of firing the 50-cal machine gun. This was a monster!!! It would shoot hundreds of yards and was capable of blowing up a body on contact. It fired tracer rounds that were red to the vision, and you could latterly "walk" the stream of rounds to the target of death!

We were also introduced to the hand grenade. We would be in pits with an instructor and given live grenades to throw over the concrete wall. It never happened to me, but the stories were rampant that the recruits would sometimes freeze with sudden fear and drop the Grenade!

We got to watch the ONTOS perform. It was a tracked vehicle that looked like a small tank. The unit had six recoilless 106 cannons attached to it. It could drive at a very fast speed, screech to a halt, use the 50 Cal machine guns that were attached to it, fire red hot rounds tracers to locate the target, and then fire one or all six guns!! It was an awesome weapon to see.

We went into a Quonset that was filled with tear gas to demonstrate and use our gas masks! It was quite a surprise when they asked you to remove your mask and sing the Marine Corps Hymn before you could get out! The puke and mucus was blowing out of all!

One great memory was a night maneuver where over 300 Marines were set on a hill. The mission called for a full fire across a valley. The killing fields, lit up with tracer rounds, looked like a walking red wall. You wouldn't think anything could escape death!!

LETTER # 27

Well, it shouldn't be too long, and I'll be there. That is, if these ole' legs will hold together for another week. I can fake it most of the time. The runs are what really screw me up. I really have a problem then. I'll be damned if I'm going back to sickbay. They will take one look and set me back in training again. NO MORE!! I can take it. I guess these stress fractures never heal up. I can feel a bump on each shin. No sweat, I will make it this time, or they will carry me out. The guys I joined with from the NODAK PLATOON are already back from their leave. (I haven't even finished training) I heard from Vacky and Vern. They sounded like they had a good time in Ole' North Dakota. They said the only problem for me was… they drank all the beer in Minot! They said Arny's had to order in a semi-load just for them! They also said my old girlfriend, Peggy, was sad, but they all consoled her for me — quite a few times!! Actually, Peggy sent me a letter and said she wanted to talk when I got home. She wants to know where we stand!

Vern told me that Sgt. Weets, the slime ball recruiter who got us in this mess, said I was the winner of both trophies, the high shooter and the

```
outstanding Marine.  He also said he got to see
the  trophies,  and  they  were  two  California
fruits,  holding  hands.  I  think  he  was  just
jealous!!!
LOVE, DALE
```

Okay! This letter brings back huge memories and guilt!

I finished ITR and was granted 20 days of leave. I caught a plane to Minot, North Dakota! My mom, brother, and neighbor met me. I traveled like many new boots do, in my uniform. I changed planes in Minneapolis, and as I was waiting, I met this young lady. We got into a deep conversation, and I found out she was from Miles City, Montana. She gave me a phone number and said, "If you are ever in Miles City, look me up."

I rode home with my brother. I told him I wanted to go pick up Phil's car. When the rest were home on leave, Phil had left his car at his girlfriend's house. I was getting in the car when Judy (a girlfriend of another of us) said, "Hi!" I talked a little, and she said there was a party over at the park. She wanted me to give her a ride! Well, long story made short, I never made it home. I partied for days in Minot, got arrested for running naked chasing a naked girl down the railroad tracks, then talked my way out of that.

After that, I ended up in Miles City, Montana. I was already over my 20 days of leave, broke, and Phil's car broke down. I finally got my recruiter to rescue me, got on a plane in Bismarck, North Dakota, and made it back to California three days late.

I reported in, and they were going to charge me with being AWOL! I explained the broken-down car and the fact that my recruiter had helped me. They settled with putting me on mess duty until I shipped out for Hawaii!

LETTER #28

> Well, I don't know if this will reach you in time Mom, but Happy Mother's Day! You are the best mom. I guess you are kinda mad at me right now. I'm really sorry about all the mess I created when I was home on leave. I really wish I could have seen more of you, but I guess I got carried away. I did see you that first day at the airport. You looked really good. I guess that's the only time I saw you. I guess I ended up in Miles City, Montana, visiting a friend and lost track of time.
>
> Please tell Peggy I'm sorry I didn't get a chance to see her. Tell anyone else who asks, I'm sorry too. Don't believe everything you hear about me when I was home. I'm sure I didn't do everything you hear!
>
> Well, how's everything back there? I'm on mess duty here, have been since I got back, and will be until I ship out. I was late reporting in.
>
> I forgot to tell you since I guess we didn't have

much time to talk. I'm going to be stationed in Hawaii.

What kind of pickup did Dad get? I guess I didn't get a chance to see it. Mom, a girl is going to bring some pictures by. I don't remember what they are about, but I suggest you don't look at them. I don't want you to give her my address, so just tell her you don't know it.

You can say, "Hi," to Peggy if you want. Tell her she can write if she wants. She probably won't want to.

I leave here about the 20th of June for Hawaii. I did find out I can take leave, so maybe I can come home then. I love you, and you are the greatest Mom ever.

LOVE, DALE

CHAPTER FOURTEEN

My time aboard the ship on the way to Hawaii was quite interesting. Obviously, being from North Dakota, I had never even seen a ship, let alone been on one. There were large groups of us that were delivered to the area where the USS George Clymer was tied up. You were put into a section, and as they called that section number, you were loaded aboard the ship. I remember the day was cloudy and drizzling rain, and I think that was an omen to mark the trip.

We were assigned a bunk, shown the chow hall, and told to stow our gear and assemble on the deck. The George Clymer was so old, it had wooden decks. The Sailors called the ship "The Greasy George." I understand now that it had to do with a crooked drive shaft that caused a slight vibration of the ship, and that somehow allowed the ship to leak oil. When I arrived topside, I was amazed that the entire deck was full of Marines! I later found out that we had close to 2,000 Marines aboard, and the ship's capacity was 1,250!

I didn't get seasick right away. I tried to stay on the deck as much as possible. The air was fresh, and the smell was better. Other than drawing guard duty a couple of times, I was left alone. I found a good place under a huge air duct, and about four of us called it home! Chow was a horror story. The mess hall was down some flights of stairs, and so many puked there, that it was better to starve. Late at night, you could go to the commissary and buy some junk food. Most of us lived on that.

I got in the habit of running around the deck. I would run, exercise, and do calisthenics to keep me occupied. It was also a nightly ritual to play cards! I was pretty good at it, and I think I came out ahead. After a few days, I was able to go down to the chow hall and eat.

The head was just forward of my bunk area. It had some showers and sinks and a large half pipe that came in one bulkhead(wall) and out the other, at an angle. It had a steady stream of saltwater running through it that carried away bowel movements! The seats were mere planks in rows!

My bunk was up about five levels. I just used it to store my gear and slept up on the deck.

The first indication of land came from the Sailors. I strained my eyes to see, but nothing. Later on in the day, I spotted this dark shadow off the front of the ship. It took another day before it was visible as land.

We slowly were able to see mountains, then shoreline, then buildings. I watched as we maneuvered into a channel. (I later identified it as Pearl Harbor). It was full of ships, docks, cranes, and warehouses. The tugs guided us into a slip on a dock. The dock was full of autos, trucks, people, and birds. They informed us that it would be a few days before we could disembark from ole' Greasy George. This didn't go over very well, as most of us had all we wanted of the Navy!

I had met a swabbie (sailor) from Montana, and I tried to talk him into bringing some booze back, as they had already been to shore. He said he didn't dare to bring it back. We settled for more card playing and started an informal Smoker (boxing matches) to keep us entertained.

LETTER #29

```
Well, here we are aboard the USS GEORGE CLYMER.
We left California yesterday. Say, have you ever
heard of this strange disease called SEAS
SICKNESS? Well, believe me, it isn't as silly as
it sounds. I was fine when we pulled out of port.
If you stay on the deck and in the fresh air, it
is better. About a day out, the swabbies played
games with oysters and other things that make
your belly try to crawl out of your body. I have
my own bucket that I carry around.

Well, here it is, Day 2 on the ship. I had guard
duty last night in the forward rope lockers. I
had to walk up and down this circular staircase
in the front of the ship. It went down several
```

decks. I started throwing up at the bottom and on the way up. I did this for six hours. I thought I was going to die. The guy who came to relieve me with the Corporal of the Guard just looked at me and the bucket and started puking. I wished him well as I listened to him dumping in his bucket!

I heard there are about 2,000 Marines here. Some swabbies said it was built to hold about 1,000. He also said it was built in World War 1, and "IF" it made this cruise, it was going into mothballs!!

LOVE, DALE

LETTER #30

Well, Days 4, 5, 6, and 7 aboard the ship have been interesting. This ship was designed and built by the same people who designed prisons, dungeons, and haunted houses. The first thing you notice is the lovely sleeping quarters available on a reservation basis only. The suite we are assigned to has several hundred beds. The beds are arranged very tastefully in rows of about 50 or more. The beds are custom-built to fit the lifestyle of the dumb and mangled. They are stacked anywhere from 5 to 10 high. The length seems to be about 5 feet. They are finished with a delicate white canvas. If you are lucky enough to be assigned a top bunk, the only way to get there is to climb the lower bunks.

The fact that many Marines are six feet tall or more makes them about a foot too long! This and the fact that the top bunk is so close to the pipes you can't turn over, means you need to make a deal with the guy in front of you, so that you can sleep feet-to-feet.

Now take into consideration the fact that the Navy has speakers everywhere on this ship. They delight in making announcements like, "NOW HEAR THIS." I now add, the final ingredient-the ship is rolling and pitching. The guy in the top bunk gets seasick. This runs through the next five bunks--You add the vomit from those five to another 100, and you can imagine the smell.

As long as I'm on this subject, the chow hall is down several decks. The troops lined up at this stairwell while the ship is pitching, and the Marines are puking. You get the picture. I have found a spot topside under a big air duct that I call home! LOVE, DALE

***

The day finally arrived! We were taken off the ship and dispersed to our ordered locations.

I was headed for KMCAS, Kaneohe Marine Corps Air Station, on the other side of the island of Oahu. I got on a bus, where many were loaded in the Cattle Cars (trucks with open-air trailers). After this, the trip over the island commenced. The road took us up over a mountain in the middle of the island (called the

PALI) and brought us to the gate. The first stop was by the Airfield, and about half got off. They proceeded to a group of baby poop yellow buildings, and the rest got out. I was assigned to an artillery outfit, (3/12), Third Battalion, 12th Marine Regiment. Many were sent to infantry locations. I walked in with about 20 others to the Headquarters Building. I was directed to the Communication Department. I was going to be a 2511 (Wireman). This entailed running communication wire between the artillery pieces (guns) and connecting them up to allow the guns to have landline hookups with the FDC (Fire Direction Control Dept).

The guns were 105 cannons. The FDC Dept plotted the direction and distance the guns would shoot. As time went on, I was re-assigned as a 2531 (radio operator) and then sent to school as a 2533 (radiotelegraph operator).

My responsibility was to travel as an FO (forward observer) to the front lines to call in a location for killing the enemy. I usually had an officer with me.

I was driving an MRC-83 Jeep (a jeep with a powerful radio mounted in it). I had crawled to the top of a hill, so that the officer could get a good look at the place we wanted to shoot. We were way out in front of our troops. He was a new Second Lieutenant.

He wasn't satisfied with my choice of location. He said, "Let's go over the other side of the valley." I informed him that wasn't a very good idea. He bristled and said, "That's an order!" I again said, "I don't think you want to do that" He really got hot! He said, "That is a direct order." (I had my mic on the radio open because

I wanted witnesses to the order). I said, "Okay" and took off across the valley. Just at that time, another F.O. team called in a fire mission. I'd just reached the bottom when the rounds started landing all around us. I got to the other side and received a message from our C.O. to immediately report back to the F.D.C. tent. I arrived and told him what I was ordered to do. The Second Lt said, "NO!" It was my fault.

I asked to have the log at the radio in the F.D.C. tent read (all incoming transmissions are logged). The C.O. took the Second Lt. out, and we never saw him again.

There were all types of training in Hawaii that should have clued me into the potential duty I might be facing. One such was POW training. It was definitely insightful, and I hoped I would never need the knowledge I learned that day.

We were taken into a jungle area and dropped off. The instructions were to find our way to a certain part of the island, using a map and compass. There were about six groups. My group was on the west side and needed to go east.

We had three days to make the trip. We were to evade the mock bad guys that were hunting us. We had no weapons, no food, just one canteen of water.

Most groups immediately set out. I held ours back. We crawled up a hill and camouflaged ourselves. We waited and watched down the hill as the mock bad guys captured a group. They tied them in a neck rope line and led them away (we were instructed NOT to fight back).

The bad guys searched the area and started up the hill,

looking for a sign. One walked within two feet of me, stood there looking around, took a leak, and went past us.

We waited for about an hour and set out east! We were suddenly surrounded. They knew where we were and were waiting. They put us in the rope line and trailed us to the POW Camp.

Large holding cells were set up. They were bamboo cages, and we were crowded in so many, that you had to stand.

They took you out, one at a time. You were taken to an interrogation room, and various methods were used to try to get you to say more than your name, rank, and serial number. The only one that got me was the "Apache Pole." It was basically a telephone pole where they wrapped your legs around the pole in a certain way and then lowered you down. You couldn't get off. It started hurting right away. This caused you to try to get off, which in turn only made it tighter. They didn't leave you on very long. It was just meant to show you how vicious humans can be to one another.

After all were interrogated, we were given classes on evasion and teamwork to evade enemy forces and then given the results of our experiment. It was interesting to hear that several had talked, even in a mock class!

LETTER # 31

```
We landed in Hawaii. I guess we are in Pearl
```

Harbor. I count lots of ships. You could see land for a long time before we landed. The swabbies were the first to see it. It seems funny to stand and watch the land start to appear.

I got in some card games the last few nights. We all tried to stay on topside, as the smell of the ship wasn't as bad. I won at Acey-Deucy, lost at 7 Card No Peek, Whores, Fours, and One-eyed Jacks wild.

They won't let us off the ship. The C.O. (commanding officer) said it might be three more days. I guess they weren't expecting ole' USS. George Clymer to make it!!

I met a swabbie from Montana. He and the rest of the swabbies have already been off the ship. I tried to talk him into bringing a couple of bottles of Ripple, Mad Dog, or T-Bird back.

The C.O. just announced that we get off here today!! The whole shipload of Marines roared so loud, we scared the civilians on the dock. I am waiting topside with my gear. I see large buses lined up and some cattle cars.

Well, it's tomorrow! We loaded up, and I went to Kaneohe Marine Corps Air Station. I will be stationed with an artillery outfit attached to the third battalion, 12th Marines, Golf Battery Comm section as a 2511/31.

The base is over the PALI, a big rock in the middle of the island. The base sits right on the ocean. The buildings are all baby poop yellow, and they all look the same.

I had a real surprise when I walked into the headquarters. There stood Phil Aus, my best friend!! I had no idea he was in Hawaii. He is in FDC. Good to have my best friend back!

LOVE, DALE

It wasn't all hard training in Hawaii. Some of the challenges were exactly what my D.I.s had beaten into me.

The Artillery Battery (group of 105 Howitzer cannons) I was attached to, had a company run challenge, putting the various departments against each other. One squad was the gunners(the personnel who loaded, aimed and fired the cannons), The FDC (Fire Direction Control) Platoon (the ones who plotted the firing information on maps and gave the directions to the gunners), and the COMM Platoon (communication-who had radios at the guns, the F.D.C. tent on the top of the hills, so they could find the best location to select the target. Then as the gunners fired bracket rounds (which means trying to get one high, one low, one right and one left of the target) the Forward Observers picked the middle. He commanded a "fire for effect barrage" from the F.D.C., which means dropping as many rounds as possible, as fast as possible, on the bad guys. Then the F.O.s had to quickly leave their position, as the bad guys easily figured out where you were directing the shells from.

We all set off in a squad formation in the famous, Marine Corps "quick time run." It was several miles. The goal was, you had to finish the run, in formation and with all members.

The gunners were the first to break formation and lose some members. The FDC and COMM were neck and neck. (we were all carrying full combat packs and rifles. I also had my pistol belt, holster, and 45). As we got to the last valley, F.D.C. lost a man, then a COMM runner started lagging. I was the Non-Commissioned Officer leading, so I dropped back and took his rifle. I handed it to my Lance Corporal, and I assumed his pack, all the time continuing to run. We made it in as the winners!

It was a wonder that Phil, my high school buddy, was assigned to the exact unit! I had no idea. We immediately began hanging out and discovering the island when we had liberty. They actually had Rodeo grounds on the Island. We spent many days there Bronc Riding, and Phil was also a Bull Rider.

This brings to mind my other buddy! Hank was in our outfit. He was from Fairview Community, Goose Creek Township, Union County, North Carolina! I never forgot that address! Well, one night, we talked Hank into going to the Rodeo grounds with us. We had a few bottles of Thunderbird wine with us. Somehow, we convinced Hank to try riding a Bronc! He had never been around horses. A few belts of T-bird later, Hank is mounted and coming out the gate on a bareback Bronc. Three jumps and Hank is sailing straight into the air! He hits the dirt, and I can hear the wind leaving his lungs way up in the gate area! The clowns go over to Hank, one grabs his belt, gives a big jerk upwards, and the air is back! Hank was so angry with us that he wouldn't get in the car to go back to the base! He caught a ride with someone else!

The world got really ugly, real fast after that. Some of us were selected to go do some training in Asia! We got our first glimpse of combat death, and we got our first exposure to what was going to be a long war in Vietnam.

***

My last assignment in the Marines was to a camp in Asia. The work was training the local populace on the artillery piece, a 105 Howitzer. The Marine crew had about nine personnel. There was one from Communication (me), one from F.D.C., one N.C.O. from the gun section, and six gunners.

We left the Asian base camp about sunrise, and the French Choppers dropped us off at the location. We trained for the day, and we uploaded in the chopper for the flight back to camp after. We were flying low over the trees, and suddenly, the chopper started to shudder and bend left.

The French co-pilot was on the radio screaming. I knew something was wrong. The next thing I remember was the pieces flying through the air. It was like someone had dumped a bucket of bolts into the cabin. The pilot was trying to guide us in at a slide. We hit the trees, and the cabin exploded with debris. We slid along the treetops and dropped into the forest. The next thing I remember was hitting a huge tree that suddenly stopped the chopper. No fire, but I could smell fuel and see sparks from the electrical. The sudden stop pushed the Marines forward into the lower bulkhead.

The first thing I heard was Gunny yelling, "Get the hell out of here!" We were slightly tilted to the left side, and the hatch was elevated. One of the Marine gunners was hurt in the pile up, and we got him out first. The rest followed. The crew cabin was slightly above us, and as I looked in, I saw the pilot was dead from being slashed by the windshield. His head was decapitated and lying on the floor in the helmet. A look to the right found the Co-Pilot/Radio Man slumped over. Gunny quickly got his K-Bar (knife) out and started cutting the harness holding the pilots in. Both were dead. We lowered them out to the gun crew, and they moved them away from the crash.

We were in a gully, and there were hills on three sides of us. We could see and hear other Choppers around us. We immediately started to move the dead and injured up to a clearing at the top of a hill. Two Choppers landed, and we motioned for the gun crew to get on one and then loaded the injured gunner. Phil and I were closest to the second chopper, so we started to help Gunny load the dead crew members. This Chopper was smaller, and we kind of had to hang on to the two dead ones! It looked crazy, seeing the helmet with the head in it, moving around on the floor.

The trip back to base camp was uneventful, and we were greeted at the landing by men to help remove the bodies. Phil and I went into the Headquarters tent. Gunny went to check on his men.

On the next day, we were quickly loaded up and taken back to Hawaii. We were interviewed and told to pack for stateside. We arrived at Treasure Island, and events moved quickly to get us discharged, even though we weren't due.

We were both put out, and life as civilians was quickly put on our plate!

EPILOGUE

Postscript

I lost track of Vacky after we got out of boot camp. Years later when I was in Denver, Colorado, I found out that he was working for the Colorado Fish & Game, so we hooked up and had a few beers and talked about fun times. We went on an elk hunting trip. Life then moved on, and so did we.

Later, I found out he was put in a healthcare facility and was diagnosed with Alzheimer's disease! I never saw him again because I was a Deputy Sheriff in San Diego at the time. Traveling was nearly impossible.

Vacky's family owned the farm that we partied at just before leaving for Marine Bootcamp! It was kind of the headquarters for the "gang."

I didn't have much contact with the 50 or more who formed the NODAK PLATOON after Bootcamp. I was reunited with

Phil in Hawaii though. He was on the helicopter that I crashed in. We separated after getting back stateside. He went to Nebraska, and I got lost in the Haight Ashbery district of San Francisco. I set off to Wyoming to start college, and since I had some time to kill before college started, I got a job with Brazil & Simms Construction.

Side note: When I applied for a job, the only thing they needed was a welder. I told them I was a welder! *I had no idea what a welder was!* They hired me and put me on the Graveyard shift. I looked at the portable welding truck like it was a spaceship! The other welder was an old drunk. He took me under his wing and showed me how to weld hard face rods on ripper teeth. I would weld, and he would sleep.

My ole' buddy Phil called me! He needed a job. I asked the foreman if he needed help. He said yes, only if the person had experience as an oiler (one who fuels and greases equipment). I told him that it was what Phil did in the Marine Corps. *Phil had no experience.*

Phil drove over from Nebraska and got the job. Again, the old welder who helped me, helped him.

Phil and I traveled up to Billings, Montana. We both were there, but again drifted apart. I didn't make contact with him until years later when he was dying.

Now my old partner, Phil, from Hawaii was one who I hooked up with. He lived in North Carolina. We drifted in and out of contact until, again, he was dying.

Remember a military person, especially a Marine, doesn't

usually talk about combat. If you have long stories from one who delves into the battles and the kills, it usually is just that, stories.

They were in stinking hell holes, where normal, civilized people never go; there were many jungles, deserts, and waters that create nightmares that claw at the very soul. They didn't choose to see this death and destruction that "normal" people would run away from. They did their duty with honor as they charged into the battlefields! It is to be expected that the servicemen and women return to the country changed!

They wear the ribbons and medals they are awarded by some politician or General that never got his or her feet wet! They were the same ones who sent them into harm's way, for profit, power, or revenge. They weren't part of the "negotiations" between greedy power brokers that fight their wars with pen and tax dollars. They didn't play on the political football field but were on the field of death.

Those medals don't bring back the team members who died, and the medals don't bring back the missing leg or arm. He was told to wear them. BUT—When the memories haunt him and drive him to drink or go insane, those same medals don't help him then or when he cries alone in the dark.

Those medals can't replace the buddies lost, the years of their life spent in POW Camps, the homeless streets of the Anthill cities, or the lonely mountains. Believe me. He would gladly trade those ribbons and awards to once again see those dead comrades face-to-face over a beer!

U.S. MARINE CORPS

The U.S. Marine Corps was created on Nov. 10, 1775. It is a combined-arms task force known for its focus on aggressiveness and the offensive. The Marines have been central in developing groundbreaking tactics for maneuver warfare; they can be credited with the development of helicopter insertion doctrine and modern amphibious assault.

There are approximately 186,000 Marines actively serving today, with another 40,000 Marines serving in the U.S. Marine Corps Reserves.

Source: Military.com/Marine-corps

ABOUT THE AUTHOR

Dale E. Dallman, Senior was born on a hot dry day in Britton, South Dakota, just off the Sisseton Indian Reservation, east of Lake Tewaukon, because that was the closest hospital. His relatives all worked for the railroad and/or farmed near Fort Ransom, Cayuga, and Rutland, North Dakota. When his father got the chance to quit farming and transfer to the railroad they moved to Minot, North Dakota where Dale attended grade school at Sunnyside and graduated high school at Minot High.

After Dale and a few of his friends got themselves into trouble towards the end of their senior year, a nice judge "suggested" that he and several of his acquaintances join the United States Marine Corps. The formation of the "NODAK PLATOON" made the timeline fit. The Marines took him to San Diego, California, Hawaii, Asia, and back to San Francisco, California. He grew up fast in the Marines. One of his buddies stuck with him after they were discharged and the two followed each other off and on through life.

His first real job after the service was in Billings, Montana with Brown & Williamson Tobacco Company as a road salesman. This led to other road salesman positions with the American Greetings Company and the Bristol Myers Drug Company. He ventured into real estate, casinos, auto sales, RV sales, which took him to Wyoming, North Dakota, and South Dakota. He became a deputy sheriff in California for a while, then traveled on to Washington, Arizona, Virginia, Germany, and Colorado, to name a few.

Today, Mr. Dallman lives in the South with his wife Cassandra Dallman, a professional photographer and singer out of Atlanta, Georgia. He enjoys and continues to take great pride in his children and their offspring.

This book is neither his first nor last written work so please check back to see upcoming books by Dale E. Dallman.

If you would like to know more about him, you may do so at any of these following sites:

https://www.dallmanproductions.com/

https://www.facebook.com/Grizzlytrack

https://www.instagram.com/dallmanproductions2024/

https://www.goodreads.com/author/show/49906767.Dale_E_Dallman

https://www.bookbub.com/authors/dale-e-dallman

www.ingramcontent.com/pod-product-compliance
Lightning Source LLC
Chambersburg PA
CBHW050734010526
44107CB00010B/845